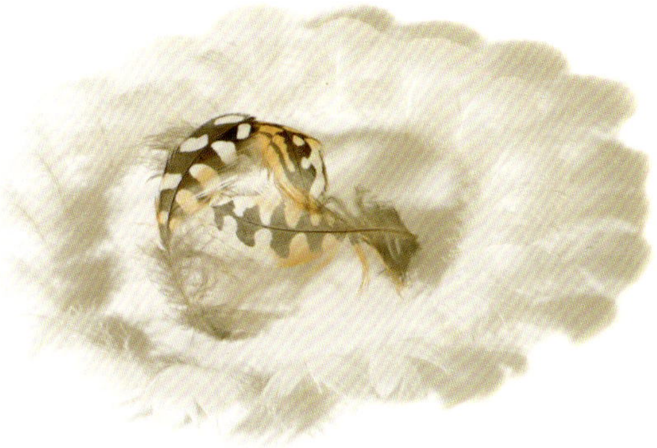

FEATHERS

Form & Function

What they are
How they work
Why we find them alluring

Chris Maynard

CHRIS MAYNARD
Author and Artist

AVIVA PUBLISHING
New York

Published by Aviva Press
Lake Placid, New York

Edited by Tyler Tichelaar
Book and jacket design by Picturia Press
Original design by Danielle Foushée

Printed in China
ISBN 978-1-940984-23-0
LOC #: 2014907301

(COVER IMAGE)

WINKY BLUE
Pigeon tail feather and
Winky Blue's shed feathers
14 x 11 inches

Winky Blue is a parakeet who was purchased by a woman when she was 75 years old to keep her company after her husband died. She picked up every feather the bird shed until she was 97 years old.

(OPPOSITE)

FAIRY WREN
(detail)
turkey tail feathers
15 x 12 inches

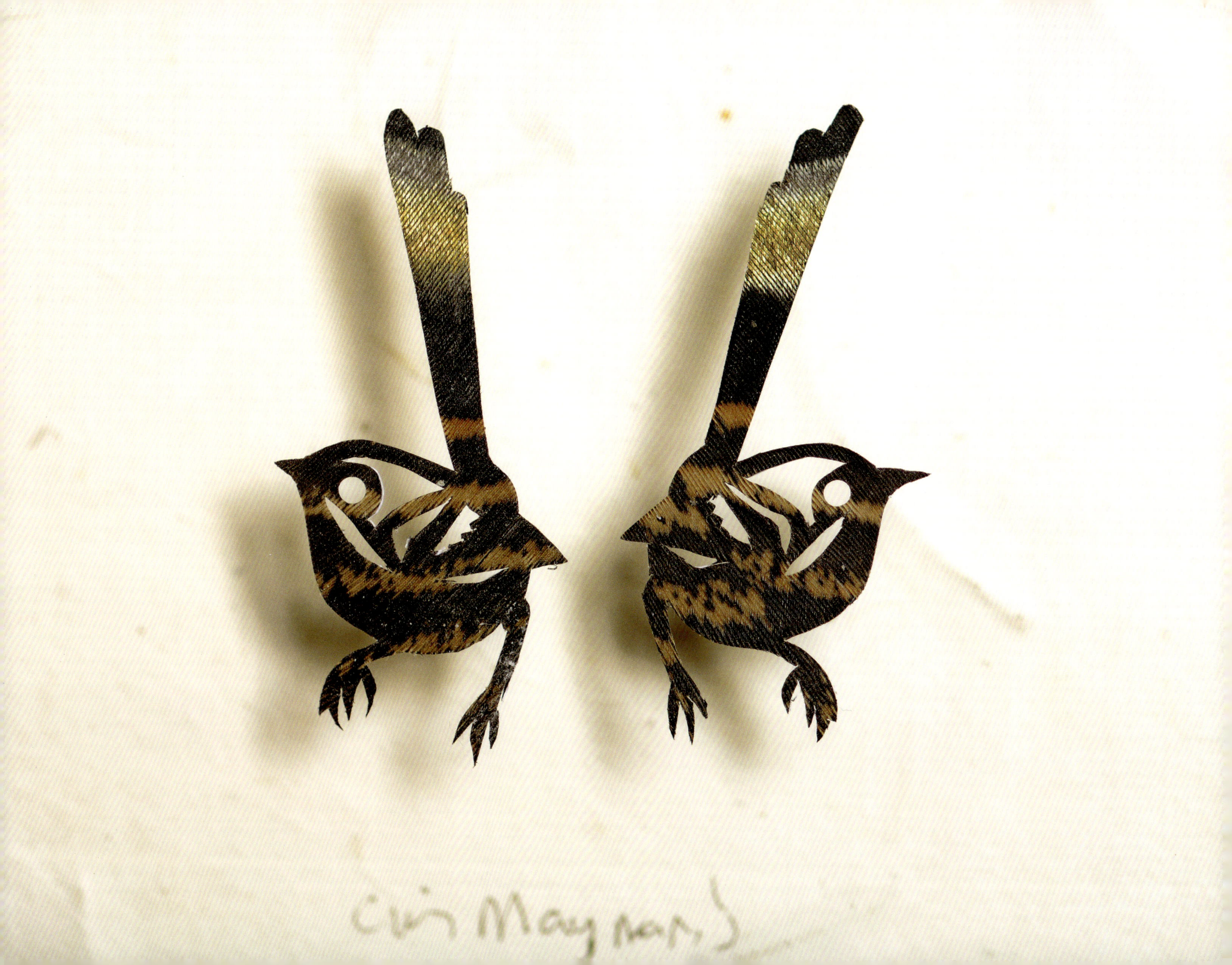
Cris Maynard

PERCHERS

Turkey tail feathers
15 x 14 inches

Tail feathers are usually best for cutting because they are relatively thin. They don't have to be thick and sturdy to carry a bird in flight like the wing flight feathers. It is difficult to achieve the detail that is found in the art of paper cutting.

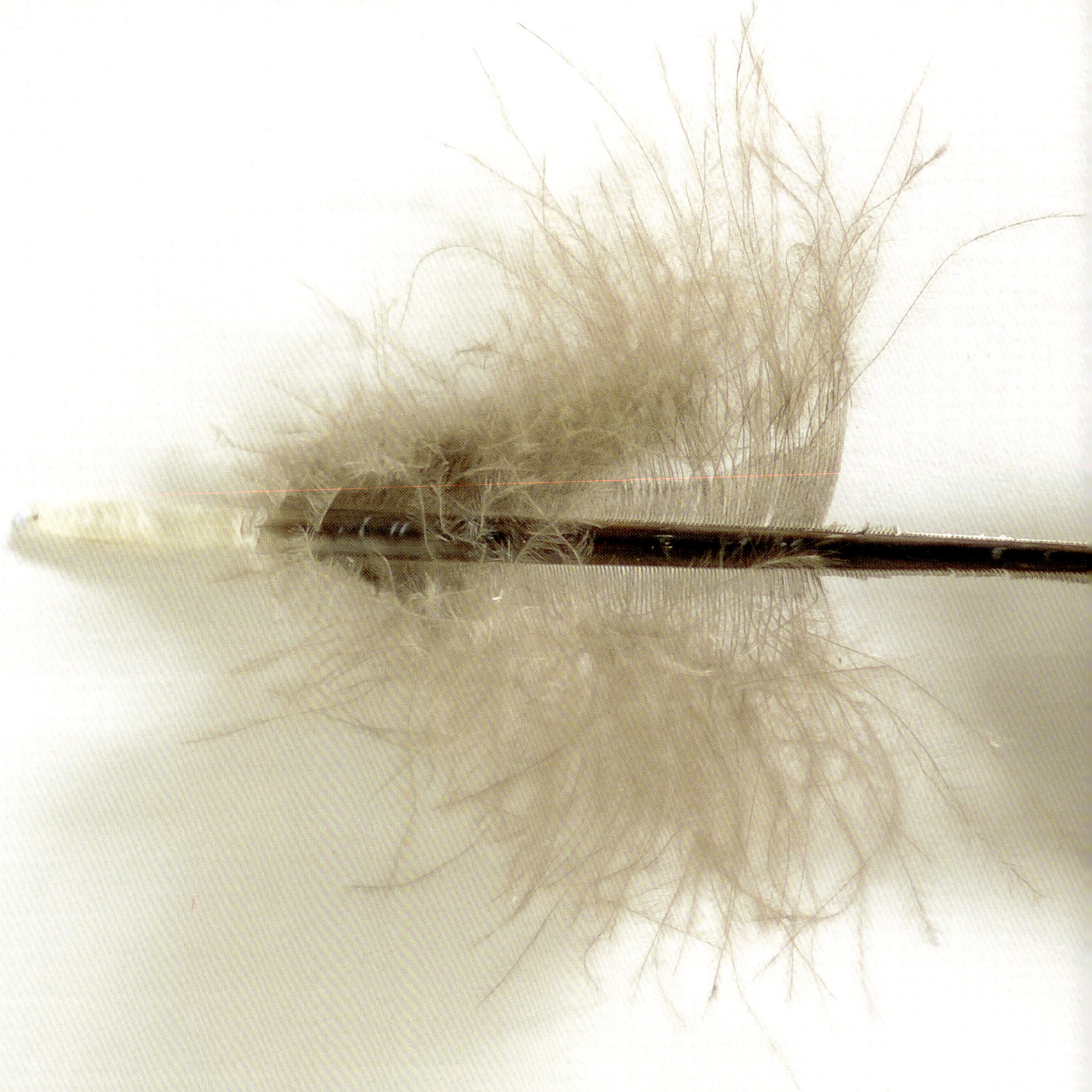

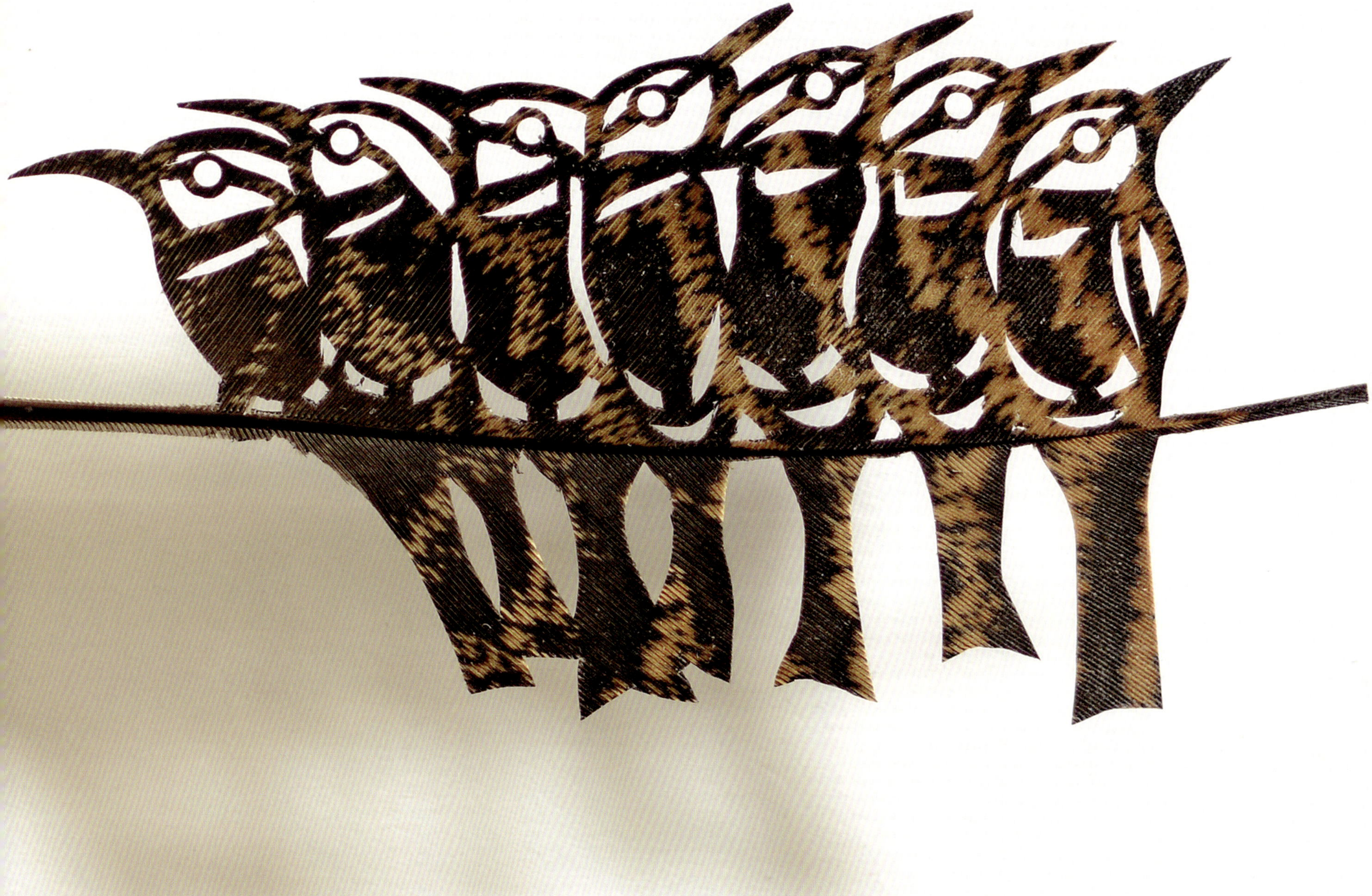

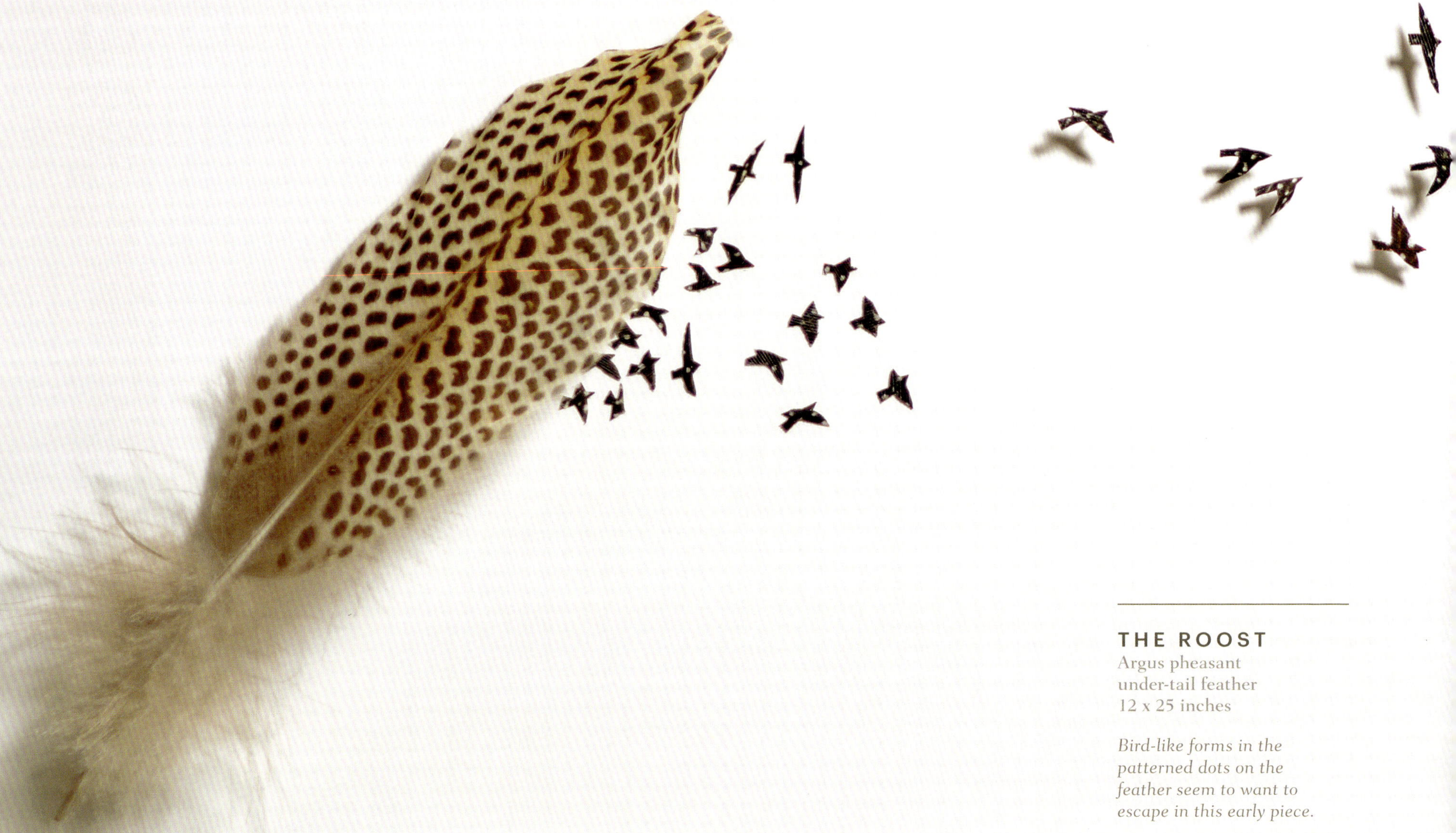

THE ROOST
Argus pheasant
under-tail feather
12 x 25 inches

Bird-like forms in the patterned dots on the feather seem to want to escape in this early piece.

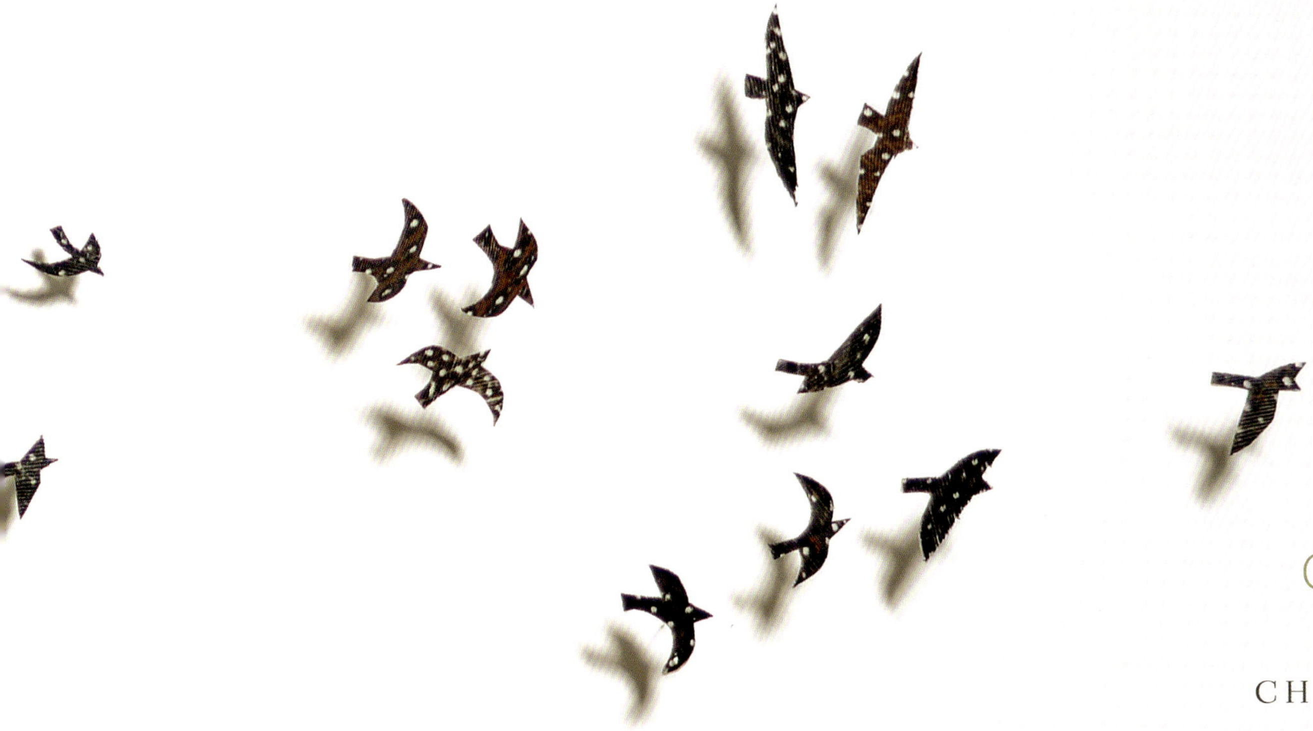

CONTENTS

Chris Maynard

HUMMERSUCKER

Red-tailed cockatoo tail feathers
17 x 14 inches

The red-breasted sapsucker drills neat rows of holes in various trees and feeds on the sap. It maintains the holes so the sap keeps flowing. Luckily for the Anna's hummingbird, this sap (which is a tree nectar) is available to slurp when flowers are out of season in the northwestern USA.

Both of these sap-loving birds show striking splashes of red feathers. The hummingbird displays metallic red on its throat, and the sapsucker presents a bright red head and breast.

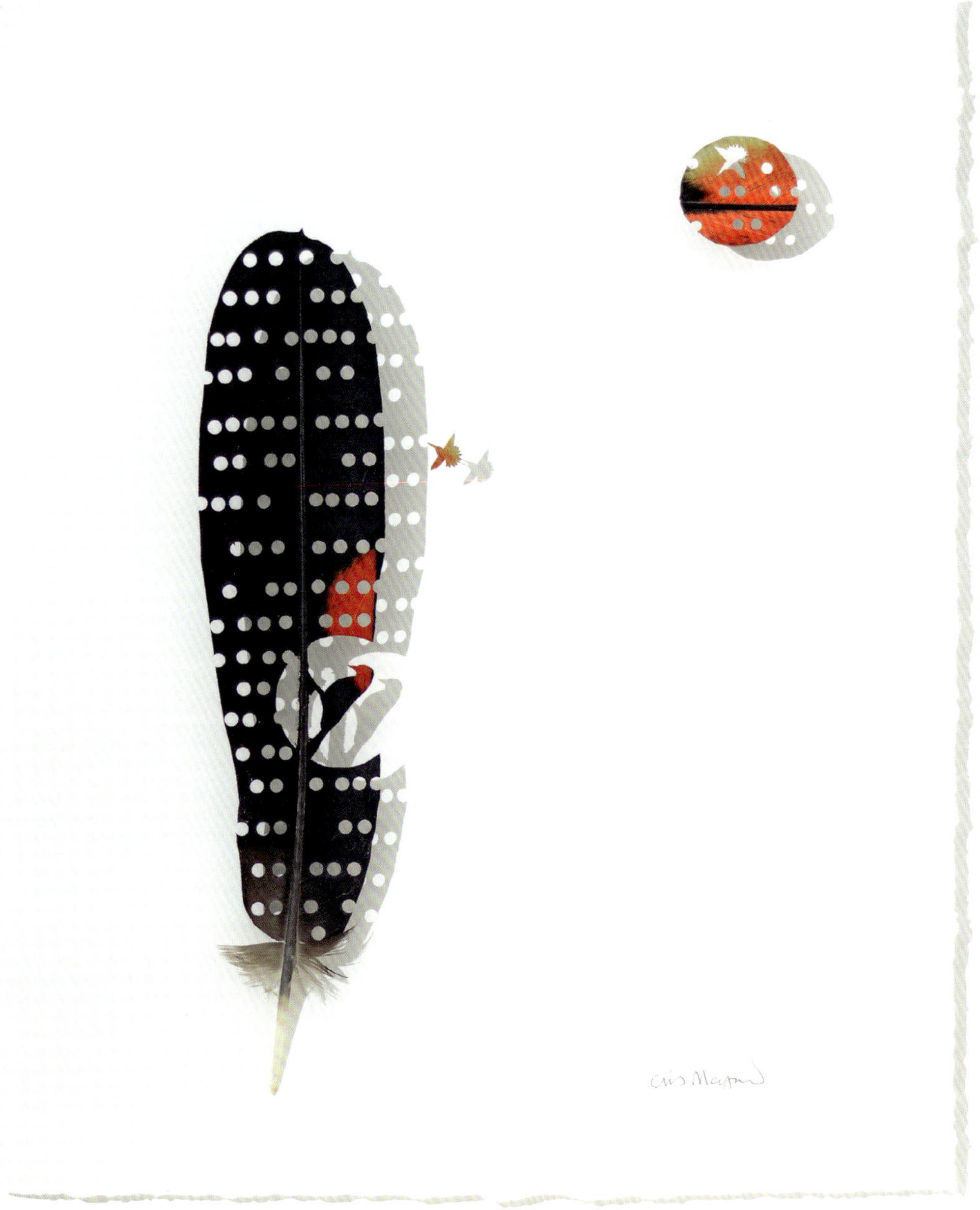

PRAISE
for feathers

Chris Maynard has taken bird art along a completely unique path with his inventive and exacting feather "carving." This book is sure to intrigue and delight birdfolk and artfolk alike.

- **Robert Bateman**, artist, Canada

Chris Maynard's work erases the line between science and art. He brings a biologist's training and a sculptor's touch to that most delicate and fascinating of media, feathers. The brilliant results are on full display in his first book, a visual feast that also explores how feathers grow, how they function, and why they inspire us all.

- **Thor Hanson,** author of *Feathers: The Evolution of a Natural Miracle,* USA

Ever since I was a child, I have been fascinated by birds and flight: feathers have always held a special attraction because of their unique and incredible design, as well as their softness, strength and beauty. As a wildlife artist myself, I strive to recreate these qualities in my paintings. When I first saw Chris Maynard's work, I was awe-struck. It requires an exceptional gift to see a picture 'inside' a feather and to be able to make that vision a reality. You get the sense that these cast-off feathers are giving birth to new life, as tiny birds take flight from a lifeless quill: the wonder of nature's endless cycle, captured in art form! Chris's pictures open your eyes afresh to the wonder of feathers and to the staggering beauty and diversity of birdlife. He combines a deep respect and understanding of natural history with an instinct that enables him to see and release the hidden potential in each feather. This beautiful book is a tribute to his passion, his skill, and his exceptional talent.

- **Colin Woolf,** artist and author of *Woodcock Pin Feather Painting,* United Kingdom

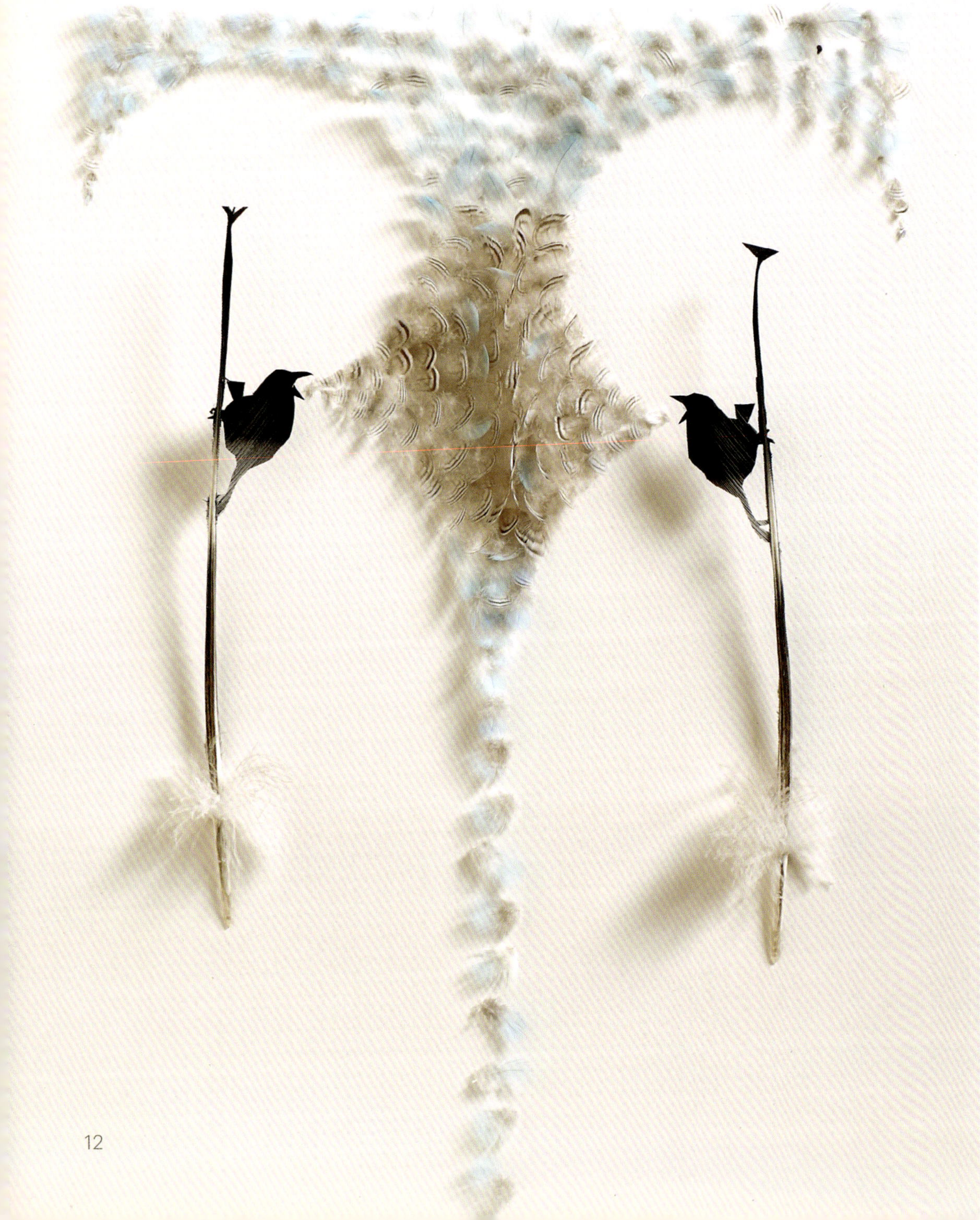

WREN SONG

Pigeon and parakeet feathers
14 x 11 inches

Pacific wrens usually live under ferns and logs. They give a tiny, short, "peep" if they are upset. And they usually are upset if you trudge around where they live. In late winter and early spring, they move to the tops of forest trees and sing—long, happy, melodious chirps. High in the trees, their songs drift like feathers to the forest floor. Their music might sound inviting to you, but it warns other males to keep their distance.

(OPPOSITE)

CHICKADEE

Pigeon feather
Unframed

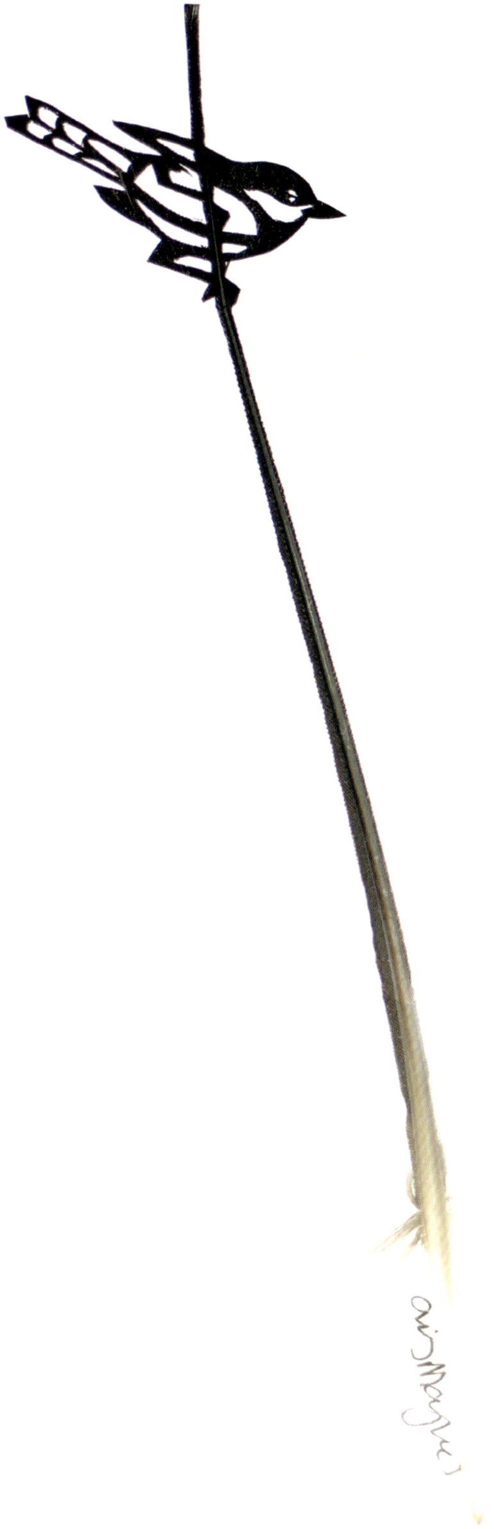

ABOUT
the author

CHRIS MAYNARD
AUTHOR & ARTIST

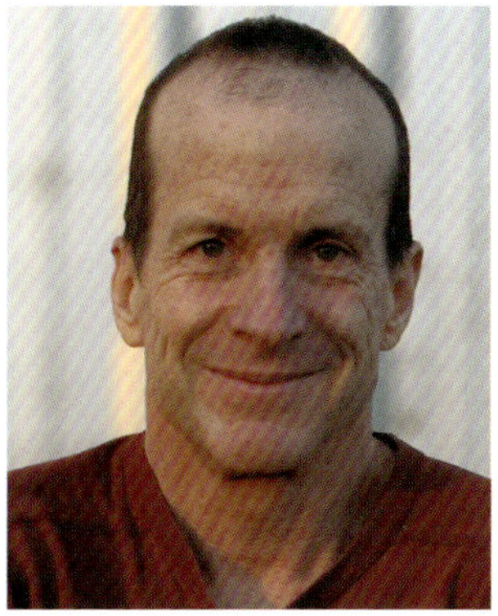

Chris Maynard uses feathers as his medium. His background in biology and his affection for life contribute to his understanding of the subjects of his art: birds. He has worked and been enamored with feathers all his life. Over many years, he developed special techniques resulting in the art featured in this book. His work has been written about, collected, and displayed around the world. He lives in Olympia, Washington

Visit *FeatherFolio.com* to purchase Maynard's prints and originals and to contact him for commissions and speaking engagements.

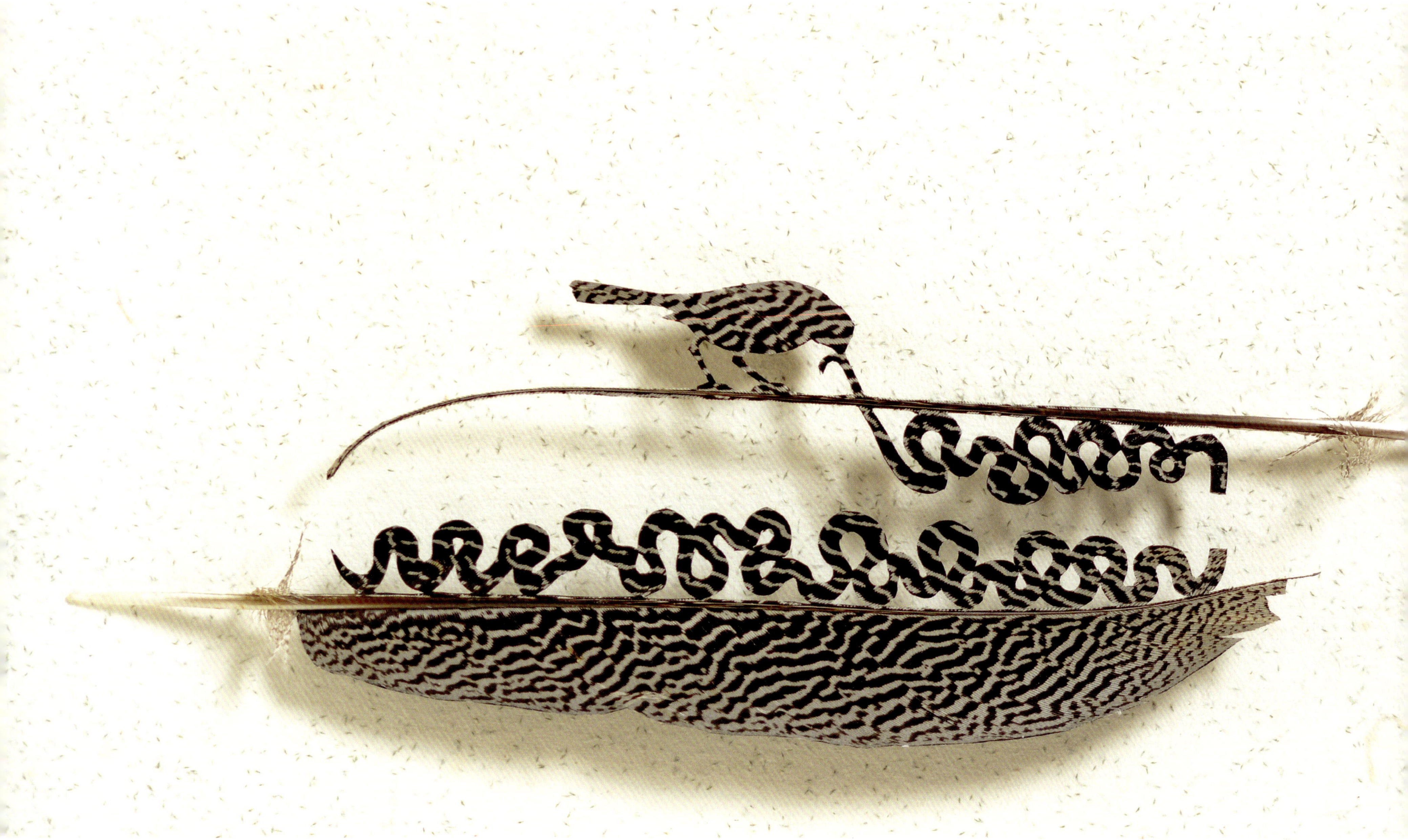

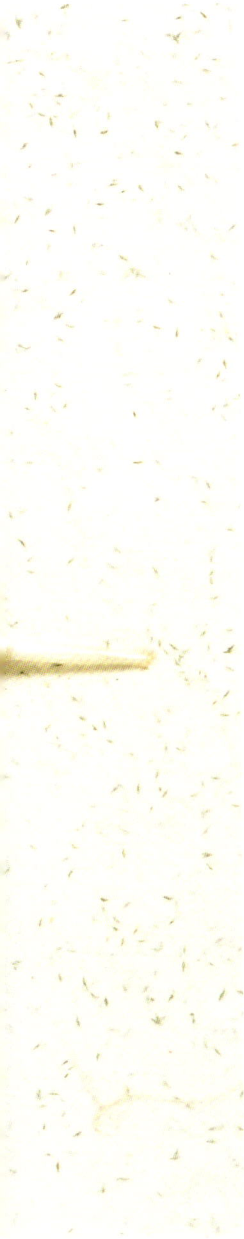

EAT 6
Silver pheasant feathers
12 x 17 inches

In Sweet Thursday, John Steinbeck describes his character, Old Jingleballicks placing one end of a clamp with a small measuring scale on his mouth and the other end of the clamp on the tip of a worm in its hole in the ground. Then he tried to pull the worm out. This was to get both a number and a feeling for how hard a Robin has to pull to free a worm. He measured one pound six ounces.

INTRODUCTION

Feathers are an ultimate achievement of the natural world. Feathers are what define birds. Other creatures lay eggs and fly, but only birds are feathered. Feathers serve many purposes, but perhaps their most notable, at least to humans, is flight: an act of which we can only dream.

Birds grow new feathers about once a year, shedding and leaving the old ones behind. This is lucky for us. It means we can delight in feathers' beauty. Have you ever found a feather intact on the sidewalk, in the grass, or on the shore of a lake? Each of these treasures carries a unique story. Maybe its owner is nearby; maybe the owner is migrating, a thousand miles away.

This mystery is what I explore in my art—the story behind feathers. A bird can have up to 25,000 feathers, but observing a feather individually gives us a different perspective of it.

As a visual artist, I attempt to capture an essence of life in an image. Because my medium is a tangible remnant of something living, I find it easier to capture this essence of life, which is so often elusive. For me, in this way, working with feathers is a direct connection between my art and the natural world.

The feathers in my art are set above backgrounds, which create shadows and make the final pieces feel three-dimensional. I find myself constantly returning to themes such as attraction, flight, and harmony. These are themes that essentially make up a bird's whole life. As you flip through these pages, keep those themes in mind, and perhaps you will begin to see feathers in new ways.

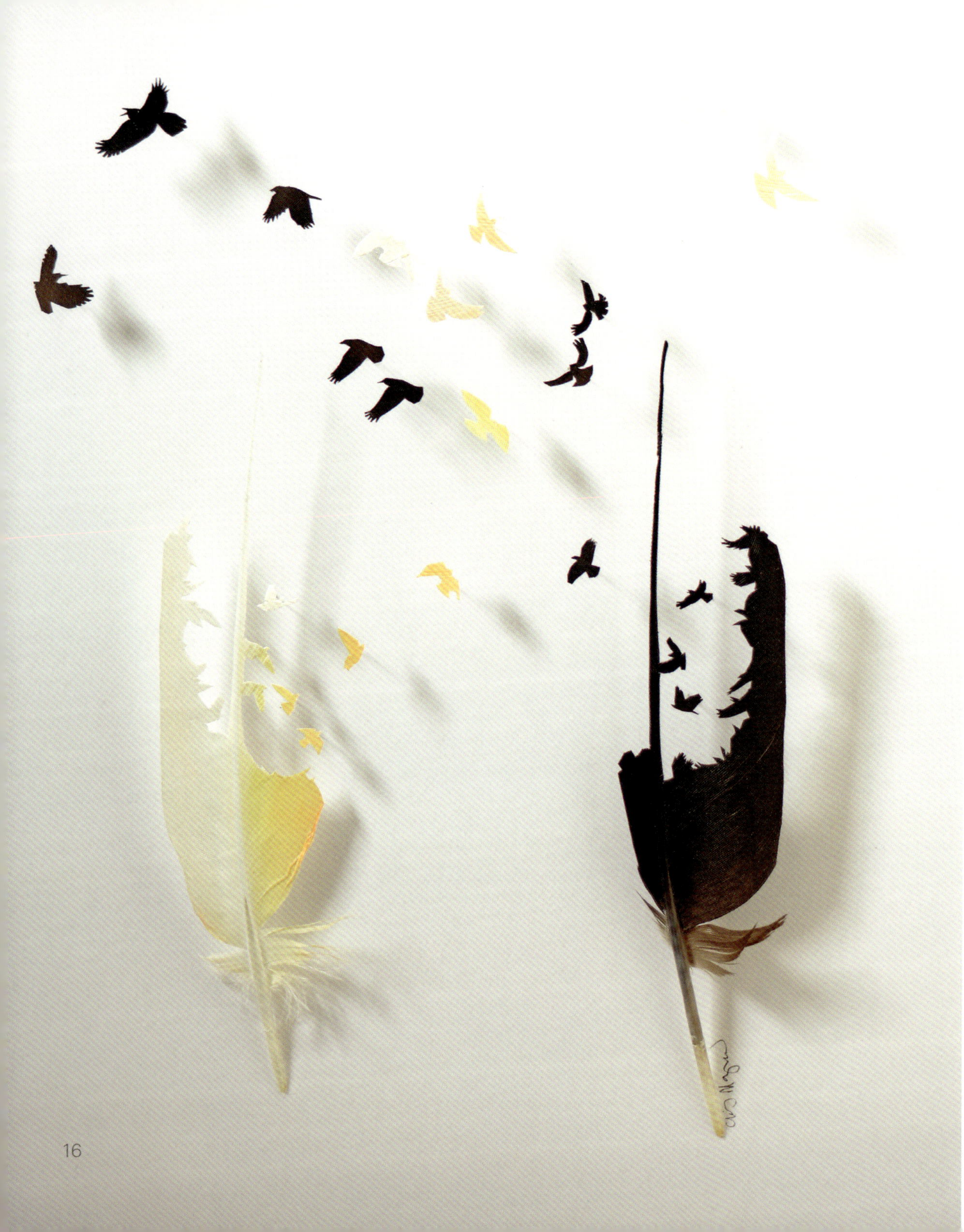

FLIGHT OF LIGHT AND DARK

Moluccan cockatoo and
red-tailed black cockatoo
wing feathers
14 x 11 inches

FORM

What is a feather?

Imagine you are holding a feather in your palm. Notice the branching form, the trunk-like shaft, and the branch-like structures called barbs. If you look closely, you may see tiny branchlets attached to the barbs—barely discernible. Feel the feather's weight. How did such a light, intricate structure come to be? How does it grow and what is it made of? When I ask myself this question, I tend to think of what happened in Washington State in 1980. Mount St. Helens erupted.

The Cascade Mountain Range runs through Washington, Oregon, and northern California. It consists of many folds in the earth's skin, pushed like the bends and crumples in the hoods of two cars that have crashed head on. The creation of a mountain range is more gradual and grand than a car collision. Among these mountains, larger and different mountains—volcanoes—arise. When the Mount St. Helens erupted in May 1980, it spewed earth's innards into a three-mile-high plume of rock and dust that settled on top of soil already made rich by previous ancient eruptions.

When I recently made a trek to the top of Mount St. Helens on a warm August day, the scramble was ashy. There were patches of dirty snow. At the top, I found a mile-wide, deep crater where the explosion took place. Its walls drop 1500 feet straight down into the earth. Standing at the crater's rim, I peered over the edge. I pictured red-hot magma pushing through a cylinder to the earth's surface, creating a huge expulsion of debris. I felt a bit queasy.

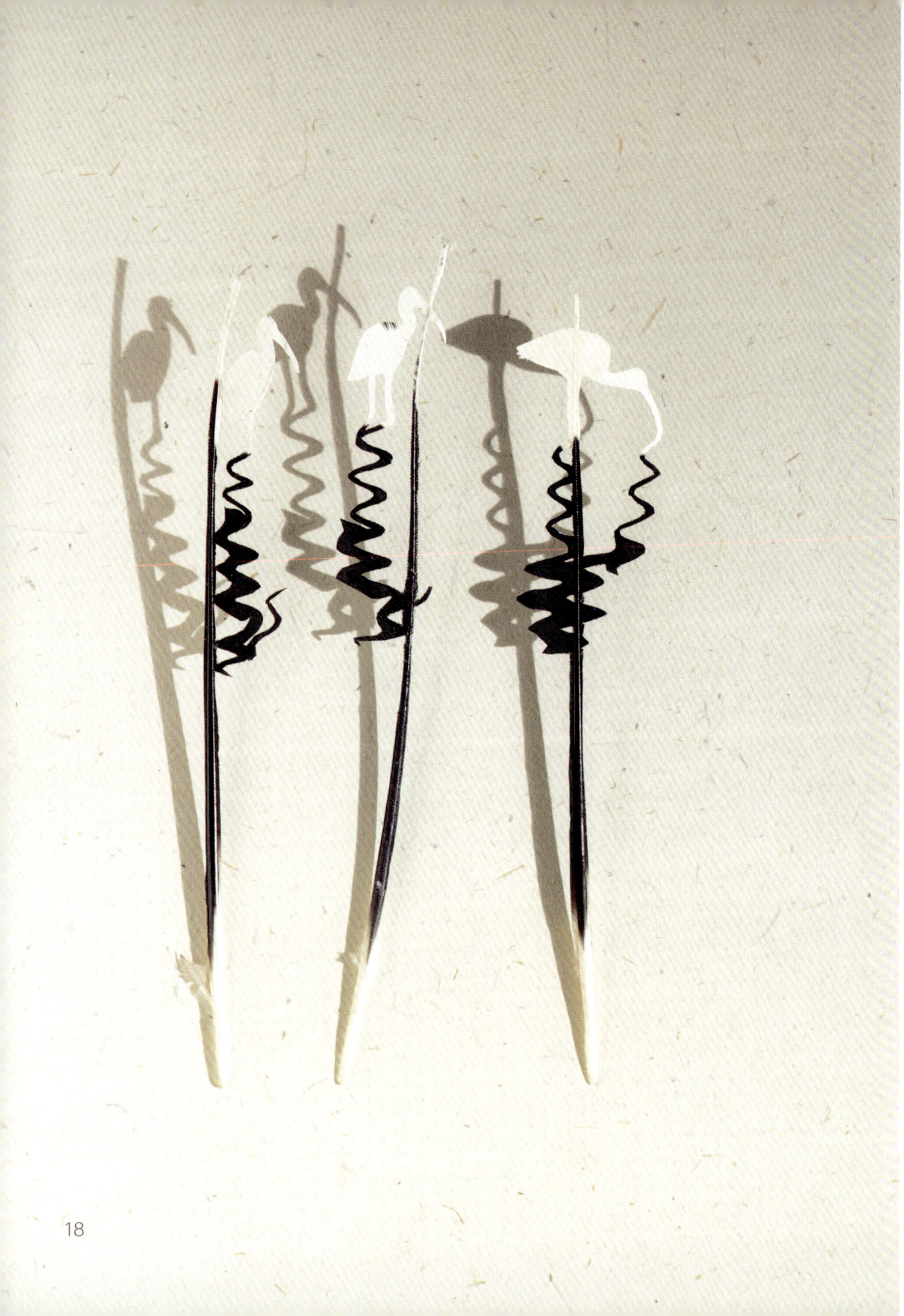

This mountain erupted from the body of the earth much as a feather erupts from the skin of a bird. Although the plume created by the eruption was uncontrolled and lasted only moments, and the growth of a bird's plume is magnificently controlled, both arise from circular tubes embedded in the skins of their respective bodies. This is the splendor of the organic creation of a feather. The "eruption," the growing process, is complexly organized and results in a lasting structure.

A volcano is a special kind of mountain, not just pushed up folds of earth. A feather, too, is formed in a distinct way. Until recently, we were taught that feathers evolved directly from scales. But no one knows exactly how feathers began their separate journey into existence. Although made of similar basic materials, feathers and reptile scales are not otherwise similar. Scales are quite simple compared to feathers. While they form through a hardening of the outer layer of the skin, feathers erupt from tubes or follicles buried in the skin. Like the center of a volcano deep in the ground, each feather begins in the skin of a bird as a round tube and has its own blood supply, its own magma.

IBIS 1
Trumpeter hornbill feathers
14 x 11 inches

PERCHERS

Turkey feathers
15 x 14 inches

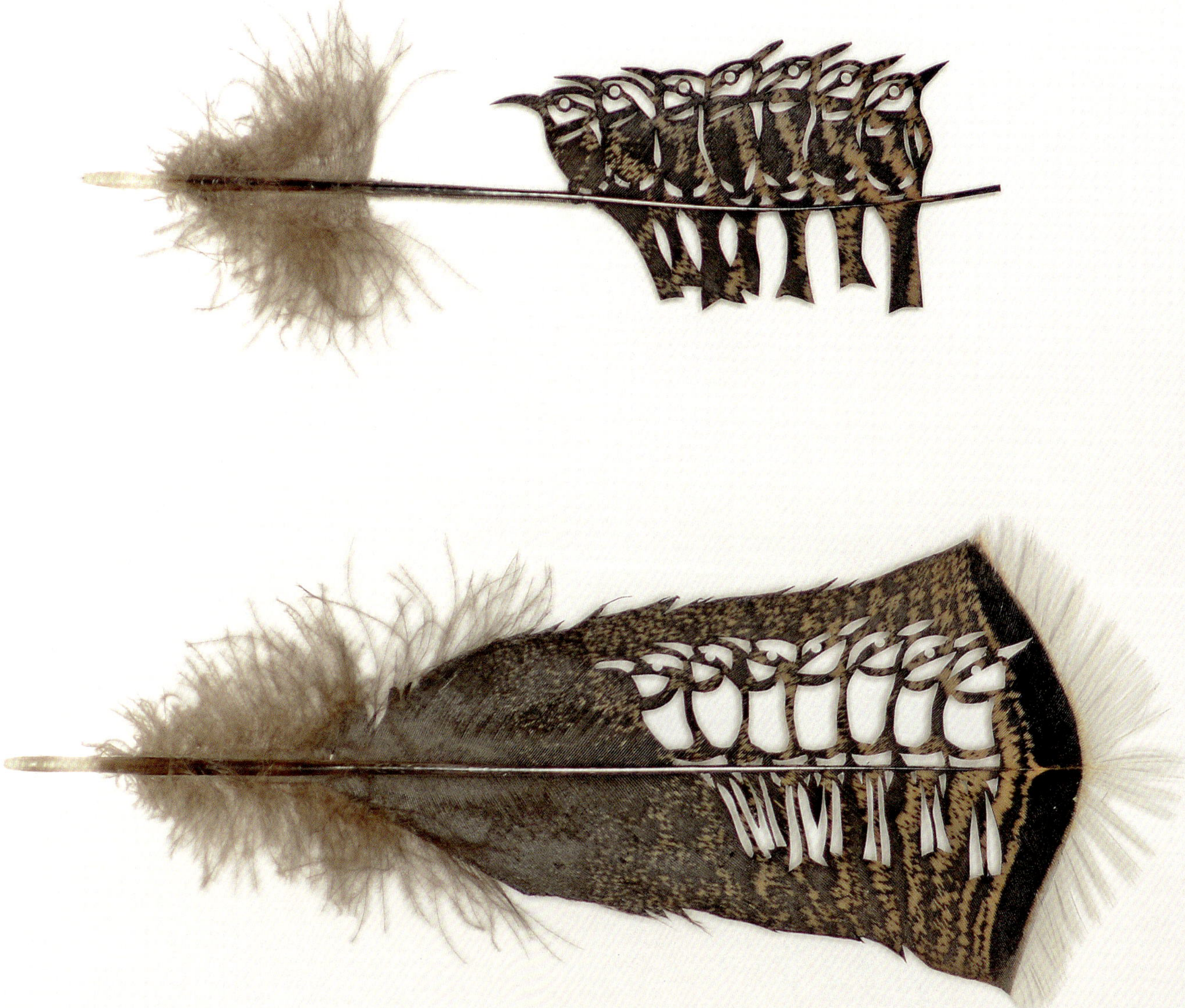

Dragon Feathers

Some dinosaurs grew small plumes
To keep them warm we assume
If dinosaurs had a few
Why then couldn't dragons too?
But dragon fire would burn them off
Unless the quills were really tough
Scales is what a dragon's got
'cause with plumes they'd be too hot

DRAGON
Turkey feathers
21 x 14 inches

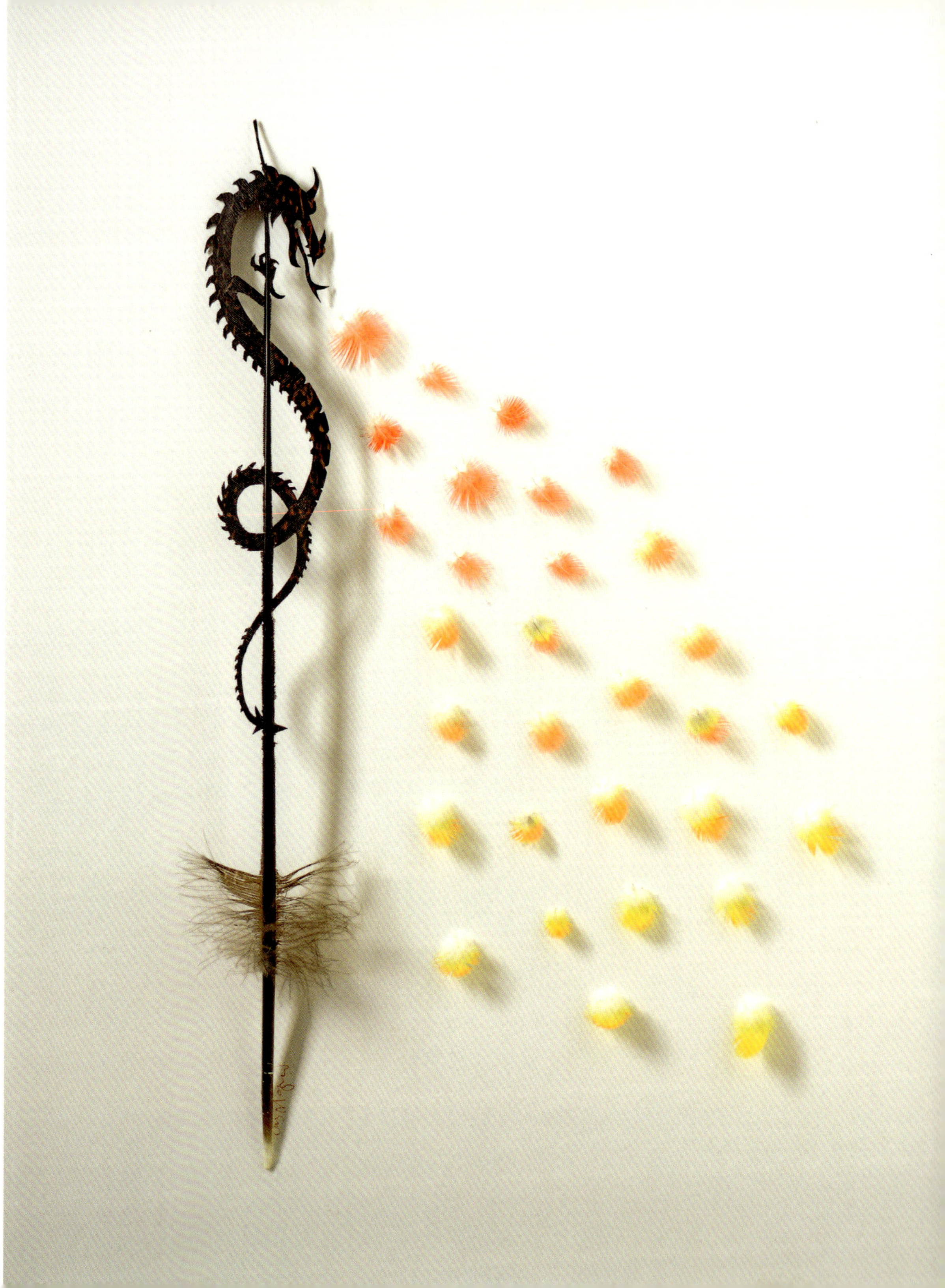

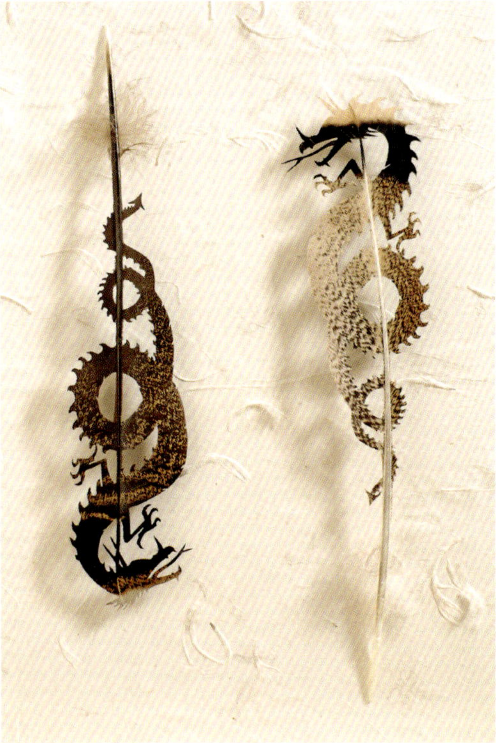

TWO DRAGONS
Two turkey tail feathers
22 x 14 inches

As birds grow new feathers each year and push out the old, the new ones are exact replicas of form and pattern.

What did the first feather look like? It probably began as a simple single filament that only later developed into larger structures with branches like a downy feather. The downy body feathers found in fossils of dinosaurs were in the process of evolving from simple to more complex structures.

Flight feathers are specialized, complicated structures with a lot of internal natural engineering. They are a culmination of successful and unsuccessful experiments over time, beginning with simple forms, like our own technological developments. Take counting for instance. First we had sticks to scratch lines in the dirt; then we used charcoal on cave walls; and we progressed from memorializing our numbers in clay to using paper and the abacus. Just in my lifetime, keeping track of numbers changed from mostly a manual affair to leaving the task to an electronic device.

If a modern day feather is this complicated, just how does it develop and grow? Each feather starts out as a little round tubular nubbin in a bird's skin complete with an artery that feeds nutrients to the cells of a growing feather and a vein that cycles the blood back through the bird's heart. The lower parts are full of blood. They are called "blood" feathers. They are alive. These growing feathers are protected by a waxy sheath.

When blood flows into a growing feather, it supplies nutrients to form new cells. This is an intricate process unlike the indiscriminate work of an erupting volcano. In his book Feathers, the field biologist Thor Hanson

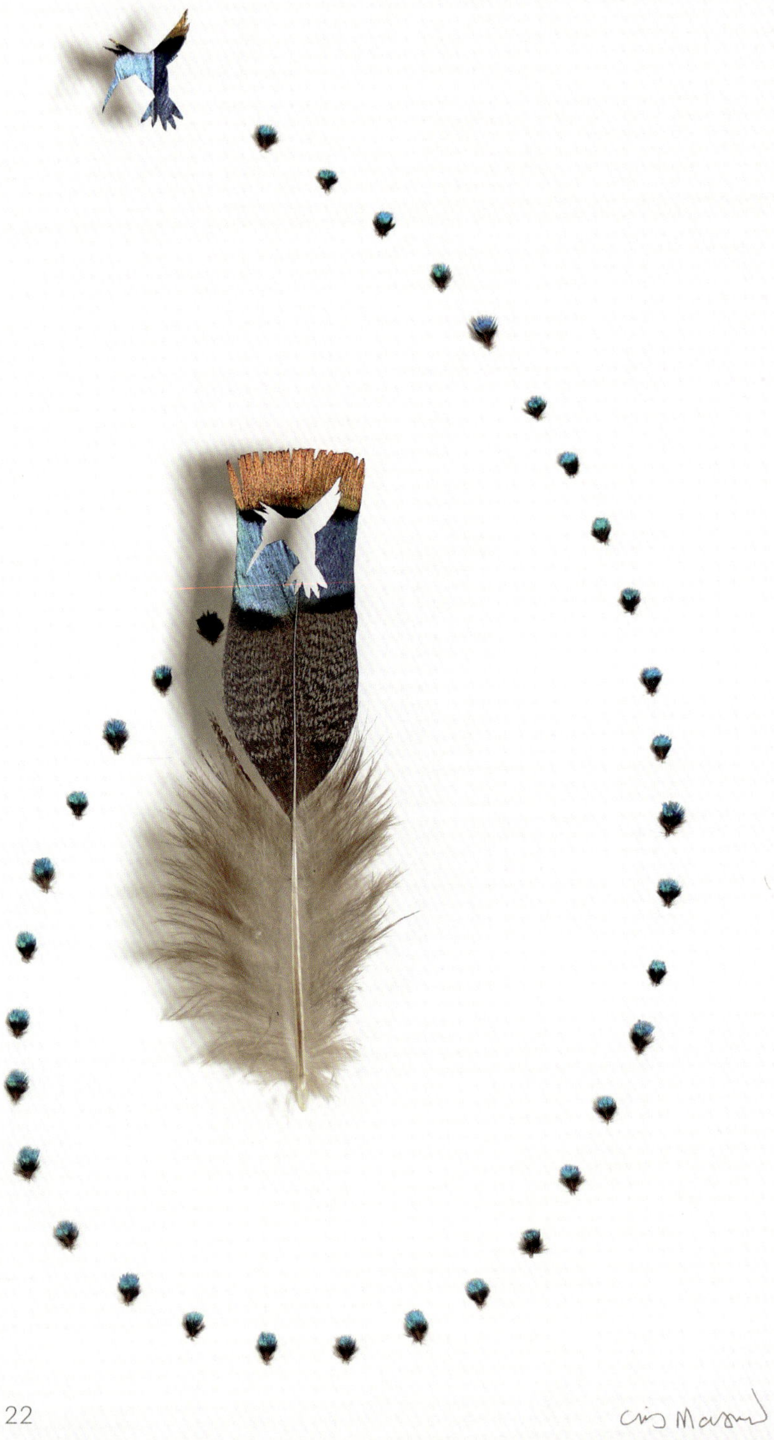

HUMMINGBIRD DANCE
Ocellated turkey feather &
tiny peacock feathers
14 x 11 inches

describes the process of cellular formation in a growing feather as being similar to "the wave," the organized stand-up cheer around a circular stadium. When people stand up to cheer, imagine them at the bottom of a round tube creating layer on layer of cells on top of each other at the base of a growing feather. And as the growing feather's stand-up cheer keeps going around the stadium, so to speak, the base of the feather forms more and more layers, pushing the formed cells up until they emerge out of a bird's skin. I like his analogy because it makes me think of feathers, of birds, of life, and of my art as one big celebration.

When the feather emerges and unfolds from its tube, the parts of the rolled up feather that faced outward in the tube become the feather's underside. The last part to unfold is the feathery down at the base of the feather. Below this base, the round quill remains buried in the bird's skin. This part is still full of blood as the feather finishes growing. While a feather is still growing, I like to call it a "feath."

My daughter Rachel and I learned something about the vulnerability of growing feathers one morning after chasing down our newly acquired domestic geese. A few moments after we released them in the field, the three adult birds flapped their wings and flew east. They looked almost wild as they took off over a neighboring house and into another field. Luckily, they were not wild birds and quickly grew tired during their first flight. We were able to catch them and carry them home. Rachel carried one in her ten-year-old arms and I carried the other two.

We decided we had to keep the birds from trying to fly away again. After getting some scissors, Rachel firmly held one bird's wing outstretched while I held it down with one hand and took the scissors in the other. The plan was to clip only the feathers on one wing of each bird to unbalance them so the bird couldn't fly. With scissors poised, we began the job. The third or fourth feather in, I clipped the soft warm shaft of a still growing flight feather. Because growing feathers don't have nerves, the bird didn't feel it, but it bled. It bled a lot. After Rachel ran to get a towel that we pressed against the feather like a wound, the blood finally stopped flowing. Though that spilled blood never got to supply nutrients for its growing "feath," this goose and all the others lived for ten years in our field. They grew new feathers, but they were content to stay with us (we fed them a lot of grain) so I never needed to clip their wings again.

As a "feath" gradually emerges and its tip becomes completely formed, it no longer needs nutrients. From the tip down, the cells die until the "feath" fully emerges. Then the artery and vein wither and dry up. When all the once-living cells are dead, the "feath" graduates into being a fully formed "feather." Only now does it begin fully to perform its numerous functions on the bird.

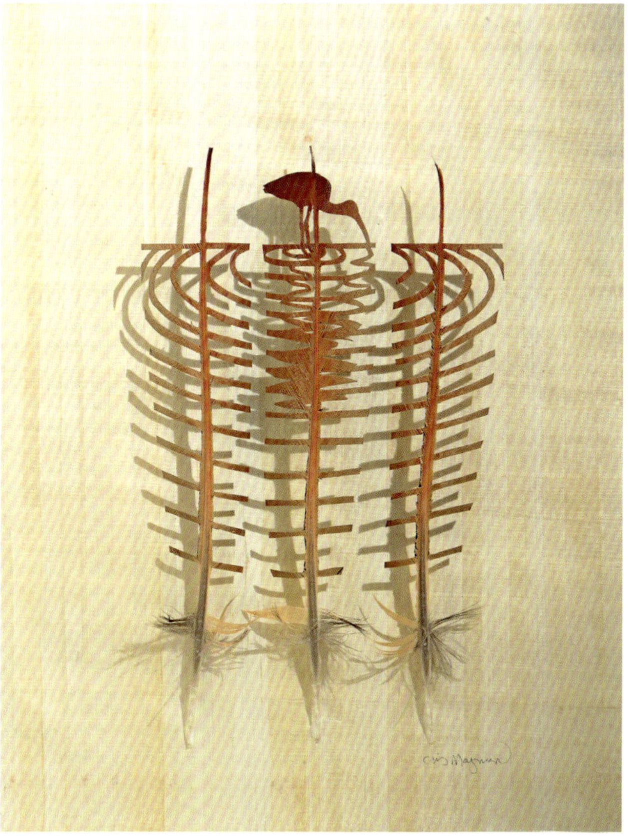

IBIS REFLECTION
Impeyan pheasant tail feathers on papyrus paper
14 x 11 inches

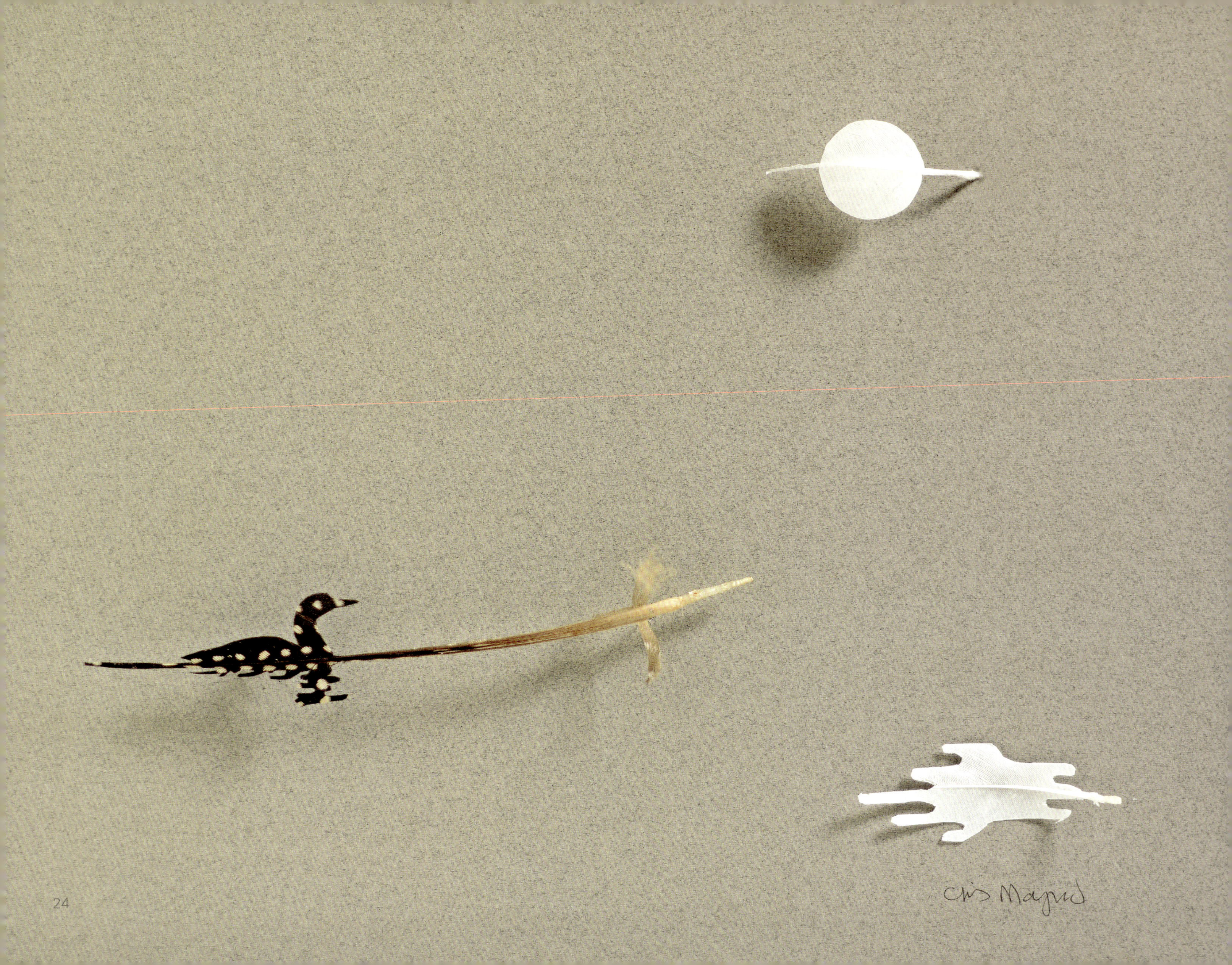
Chris Maynard

SWAN RIPPLES
Mute swan wing feathers
14 x 17 inches

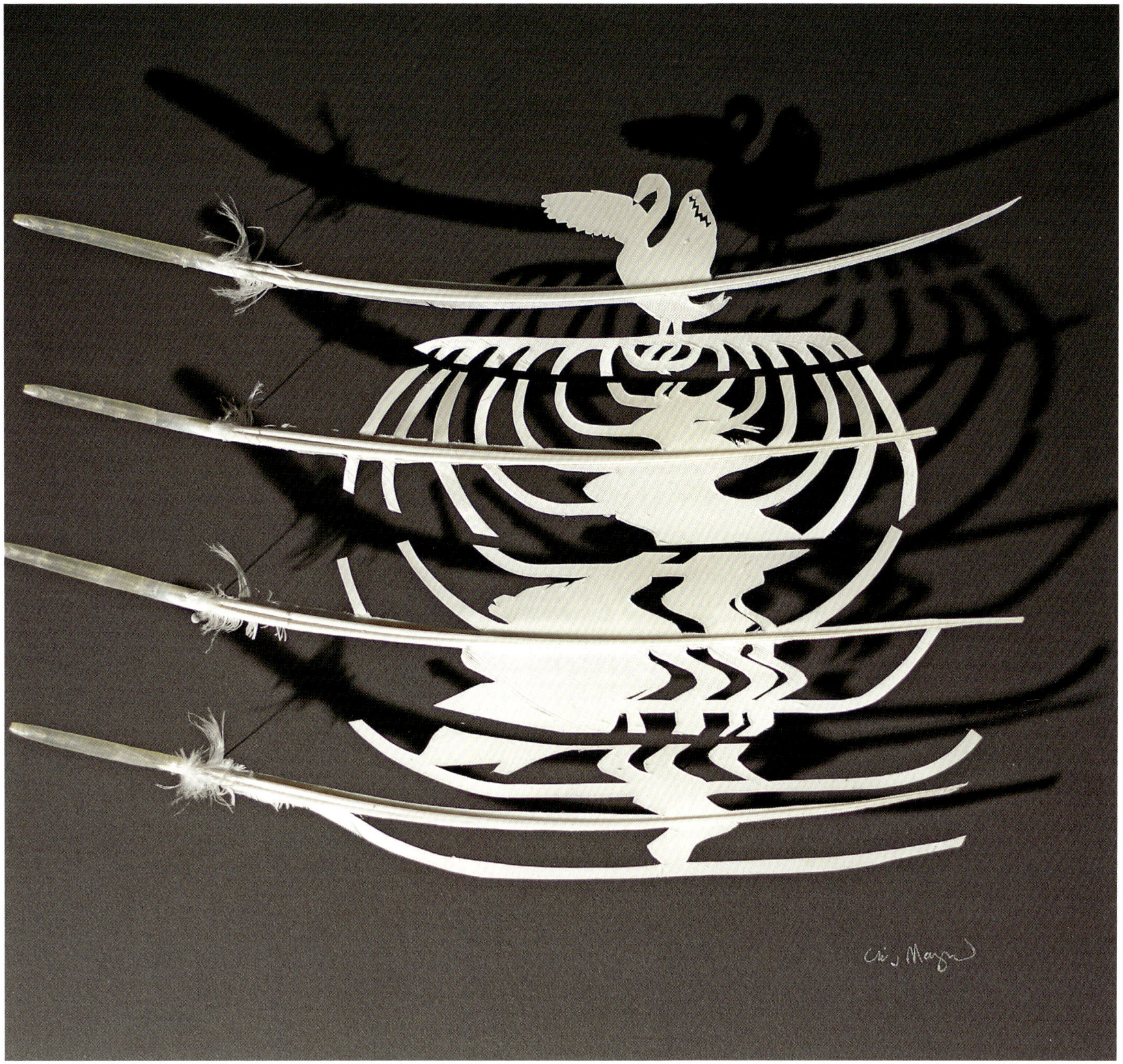

(OPPOSITE)

LOON
Guinea hen feather and
mute swan feathers
11 x 14 inches

I often marvel at the number of complicated processes and patterns occurring in feathers. As birds grow new feathers each year and push out the old, the new ones are exact replicas of form and pattern. Similarly, a bird's feathers are symmetrical on each side. When in my work I use these matched feathers, I wonder, "How is it even possible for the intricate patterns in two feathers, one from each wing, to be identical opposites?" The pigments and structure are laid down piece by piece in the base of the blood feather as cells are formed. Genetics and other processes are involved. But even knowing a little about how it happens doesn't lessen the wonder of it.

What this remarkable process of growth has created are structures made of long chains of fingernail-type protein molecules called keratin. Our fingernails, skin, and hair are made of keratin as are the claws, skin, hair, and beaks of all mammals and birds. Keratin proteins come in many forms, but there is one thing they all have in common: fingernails, skin, hair, claws, beaks, and feathers are tough.

Museum staff had scheduled one of my feather shadowboxes for a tour to several other museums around the country. When they decided not to take the feather art on the tour, I asked why. The curators told me they were concerned the feathers would break; they seemed too delicate. This is a common misperception. We often regard feathers as endearingly wispy and fragile. I speculate that this is because the keratin proteins in feathers are arranged to be as light as possible. They weigh almost nothing, and we tend to mistake this quality for fragility.

The keratin protein molecules that make up feathers are tiny, long, and spiraled. These spirals end up sandwiched together in different ways on different parts of the feather. They take the same

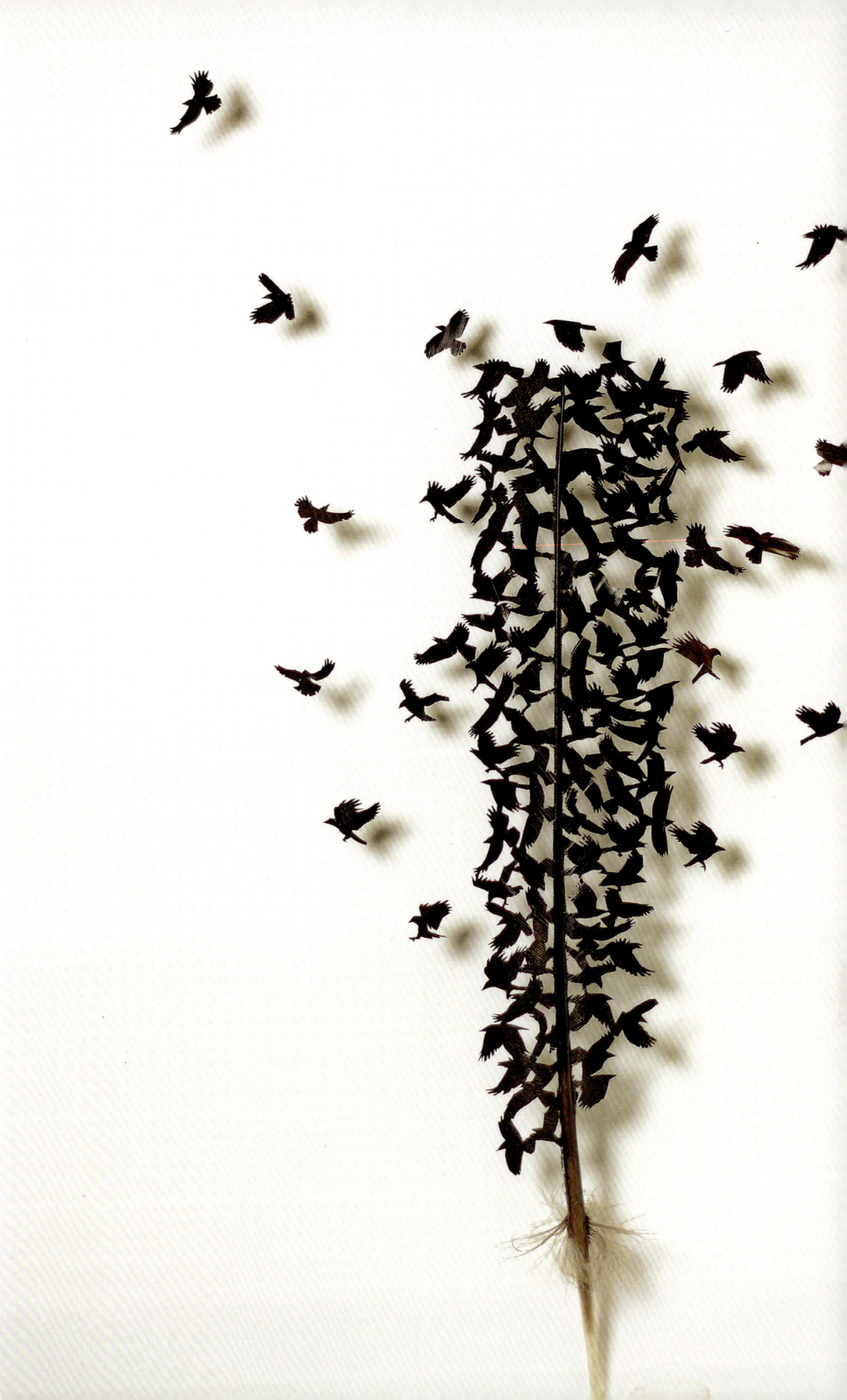

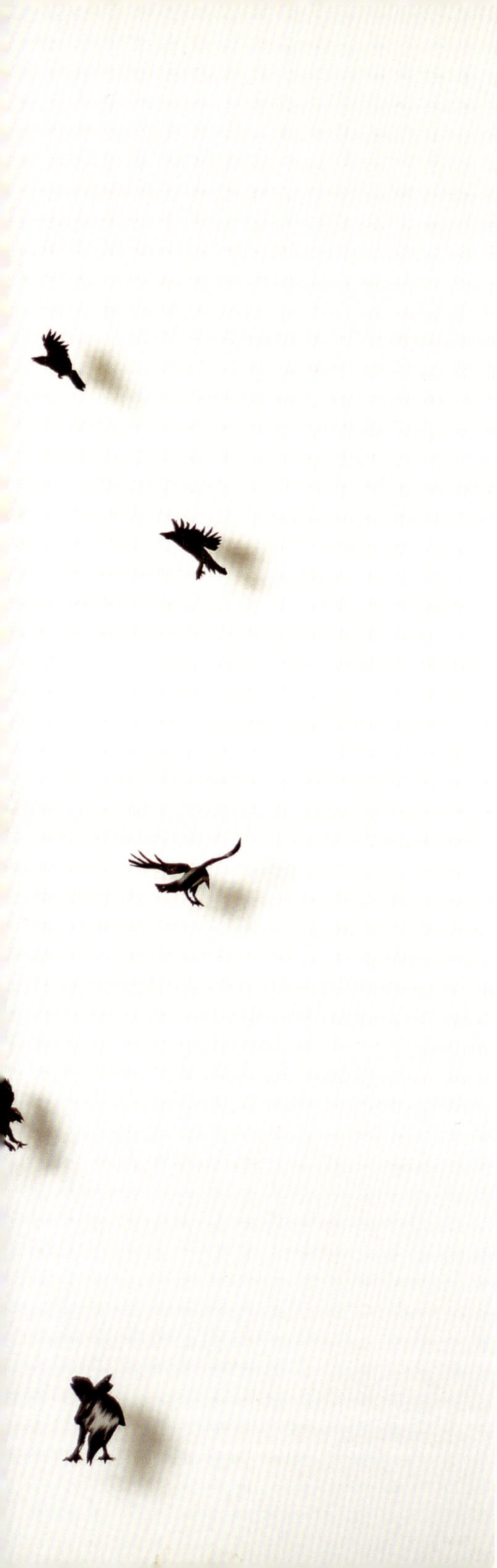

CROW TREE PERCH
Capercaillie grouse feather
22 x 19 inches

Look at a tree and you expect to see a tree. But if you stop still for a while, quietly watching, birds begin moving and the tree becomes not just a tree but home for sometimes immense numbers of birds.

form in the wispy downy parts as they do in the hard shaft except that in the shaft, they are all bundled and glued together more compactly to form the feather's hard and stiff bases.

Altogether, keratin creates the building blocks for the different parts of a feather that we know. The base of the feather we see is round and hollow (and often semi-transparent) once it is empty of the blood that supplied it. Viewed in cross-section where the feathers begin branching, this shaft reveals itself as rectangular in shape as well as filled with light, white pith all the way to the tip.

Barbs arise from its central shaft like branches on a tree. Smaller branches called barbules come off the barbs. The barbules have thousands of little hooks that keep all the barbs together. The hooks can come apart and then zip back together again like Velcro. This creates the vane, the flat surface part of the feather.

The parts of the feather that aren't all zipped together in a vane are missing the hooks on the barbules. Absent these, they don't stay fastened together and are separate from one another. We know these as down feathers, the wispy feathers of the ostrich, or the plumes of an egret.

For all birds, feathery down is underneath their body feathers. In addition to feathers of pure down, most feathers have some down, on the lower end of the shaft toward the body. Most body feathers have "aftershafts," a smaller downy feather that grows on the inside base of the larger feather. It's like a little sprout.

All of these qualities make feathers complex, long-lasting, and beautiful creations. Consequently, they are useful in art, but more importantly, indispensable for their owners, the birds.

RED TURACO DANCE

Turaco wing feather
14 x 11 inches

The pigment making this red color is copper-based only because the areas in Africa where these birds live have a lot of copper in the soil, which the plants use to make their fruit which these birds then eat.

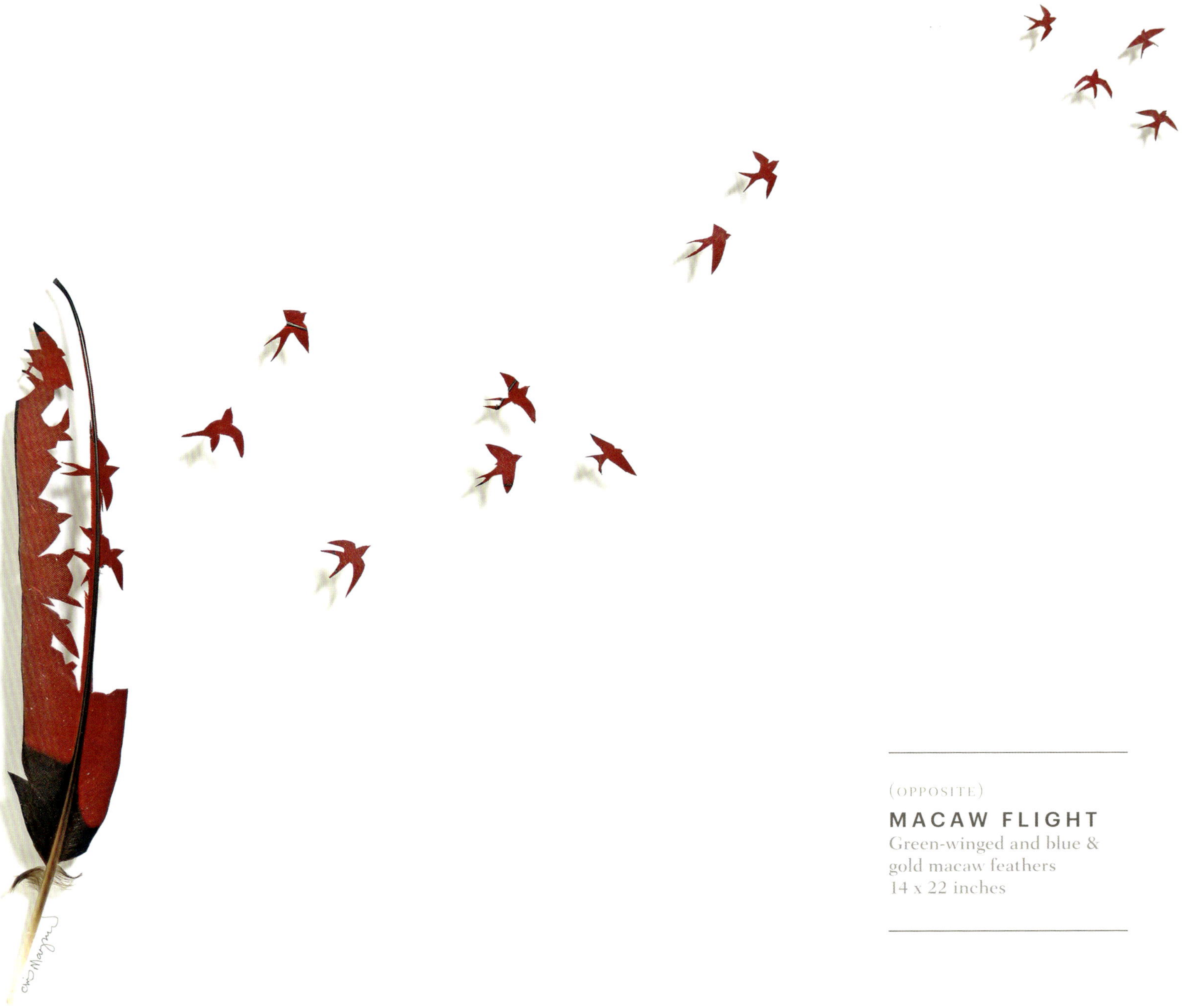

(OPPOSITE)

MACAW FLIGHT

Green-winged and blue & gold macaw feathers
14 x 22 inches

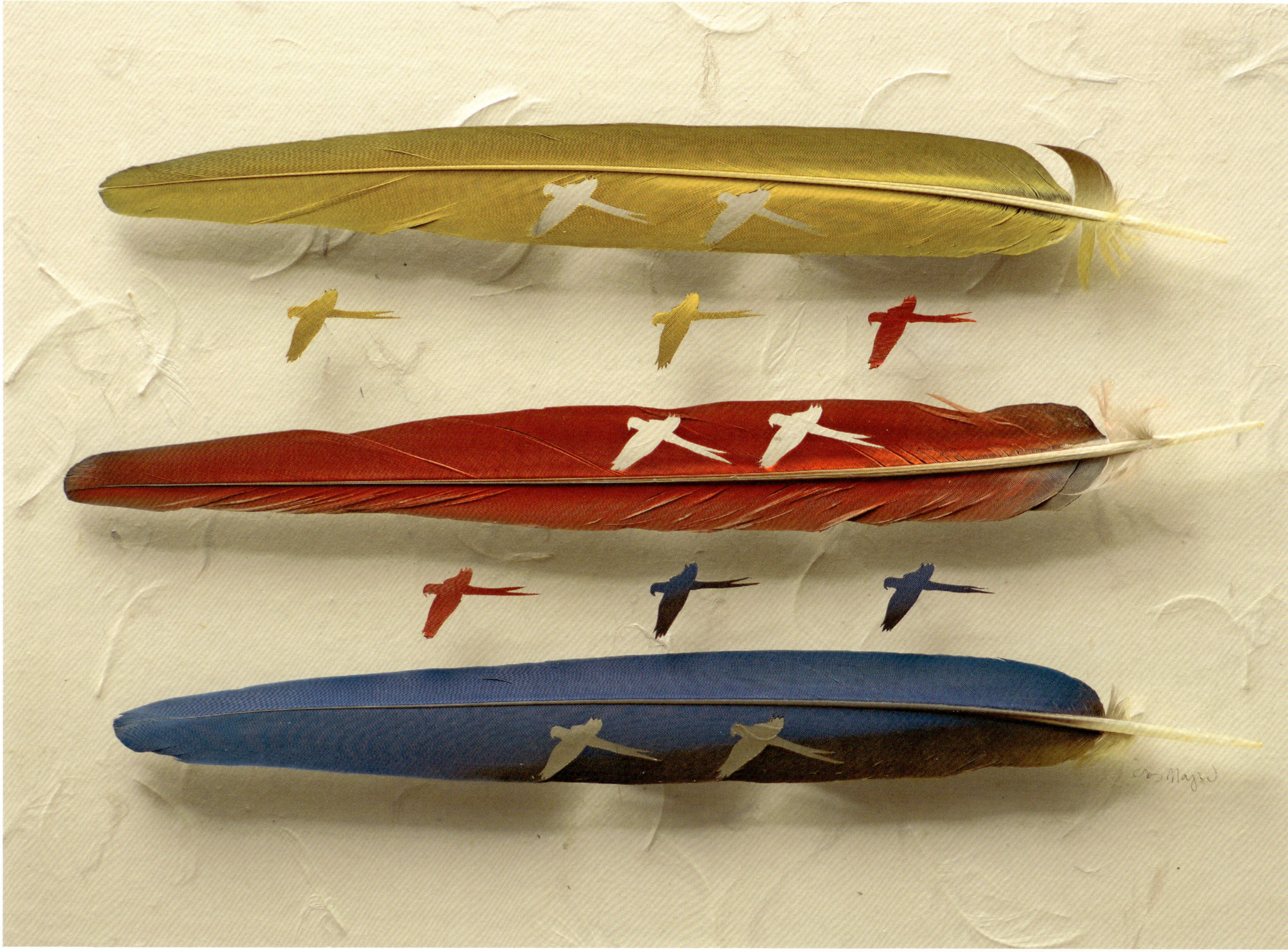

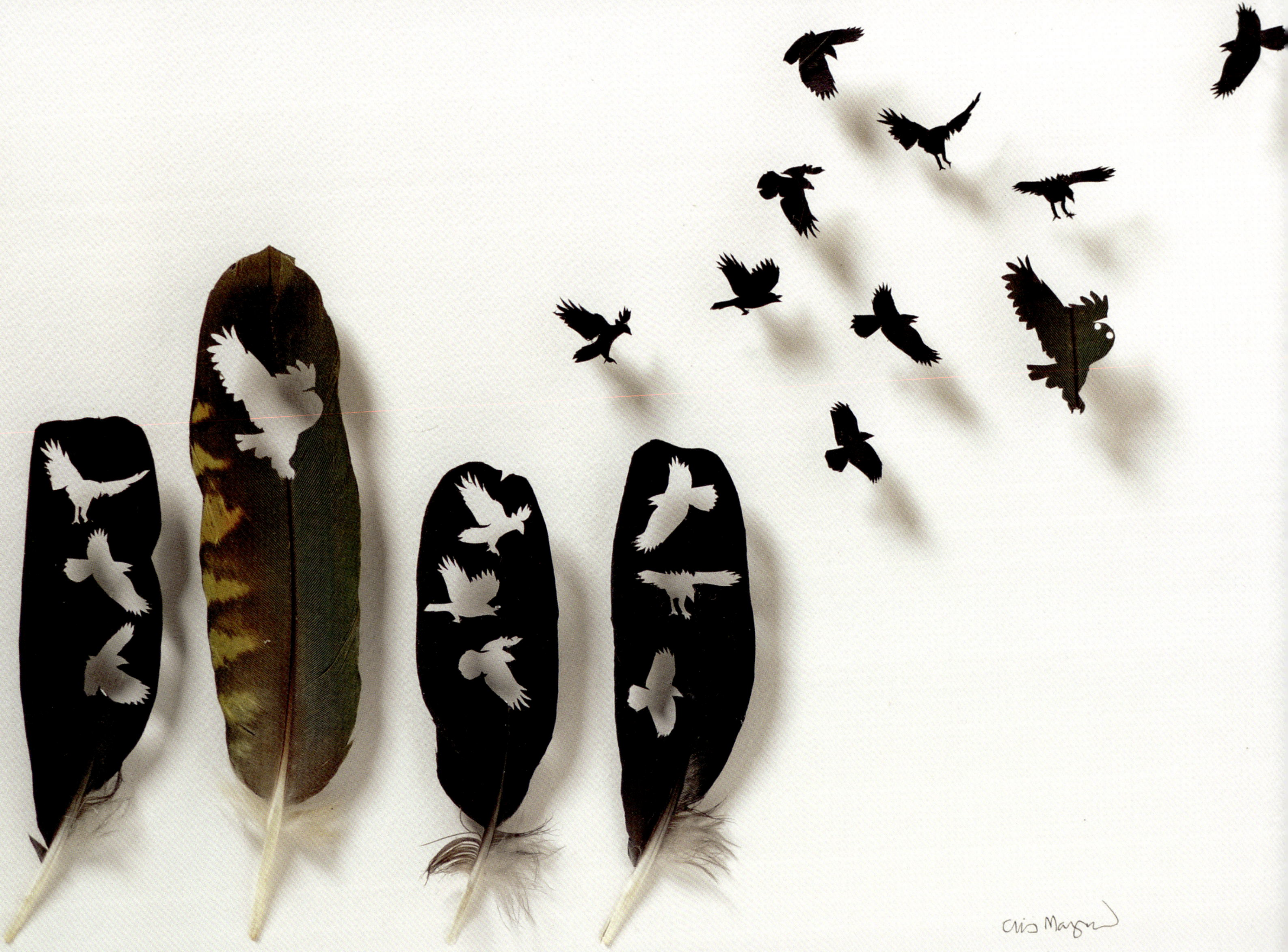

OWL FOR ONE AND ONE FOR OWL
European crow feather and
kea wing feather
14 x 17 inches

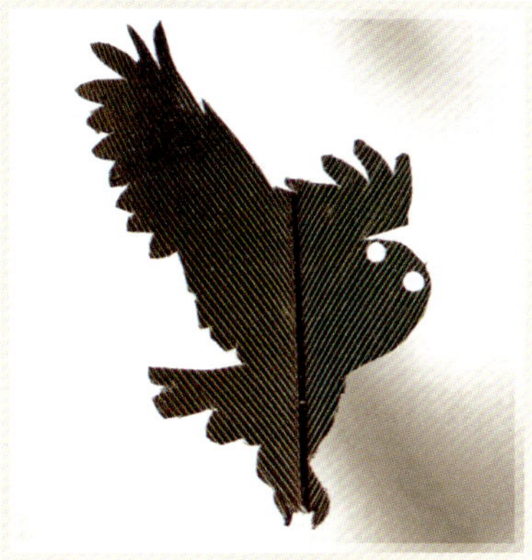

FUNCTION

How are feathers used?

Birds have anywhere from just under 1,000 to 25,000 feathers. They come in all sizes and shapes from barely visible, tiny oval cheek feathers on a hummingbird to massive tail feathers over five feet long found on a crested argus, a kind of pheasant that lives in the mountains of Laos and Vietnam. But the longest feather is from a chicken.

For hundreds of years, the Japanese bred chickens to encourage some of the typical qualities we like in a fowl: meat and eggs. They also bred them for less usual traits such as the quality of the rooster's crow and for the length of their feathers. Called onogadori, or the honorable fowl, this long-tailed chicken variety is revered as a cultural icon. Unlike other birds, these roosters don't shed their tail feathers, so they keep growing. The chickens' owners protect the treasured tail feathers by

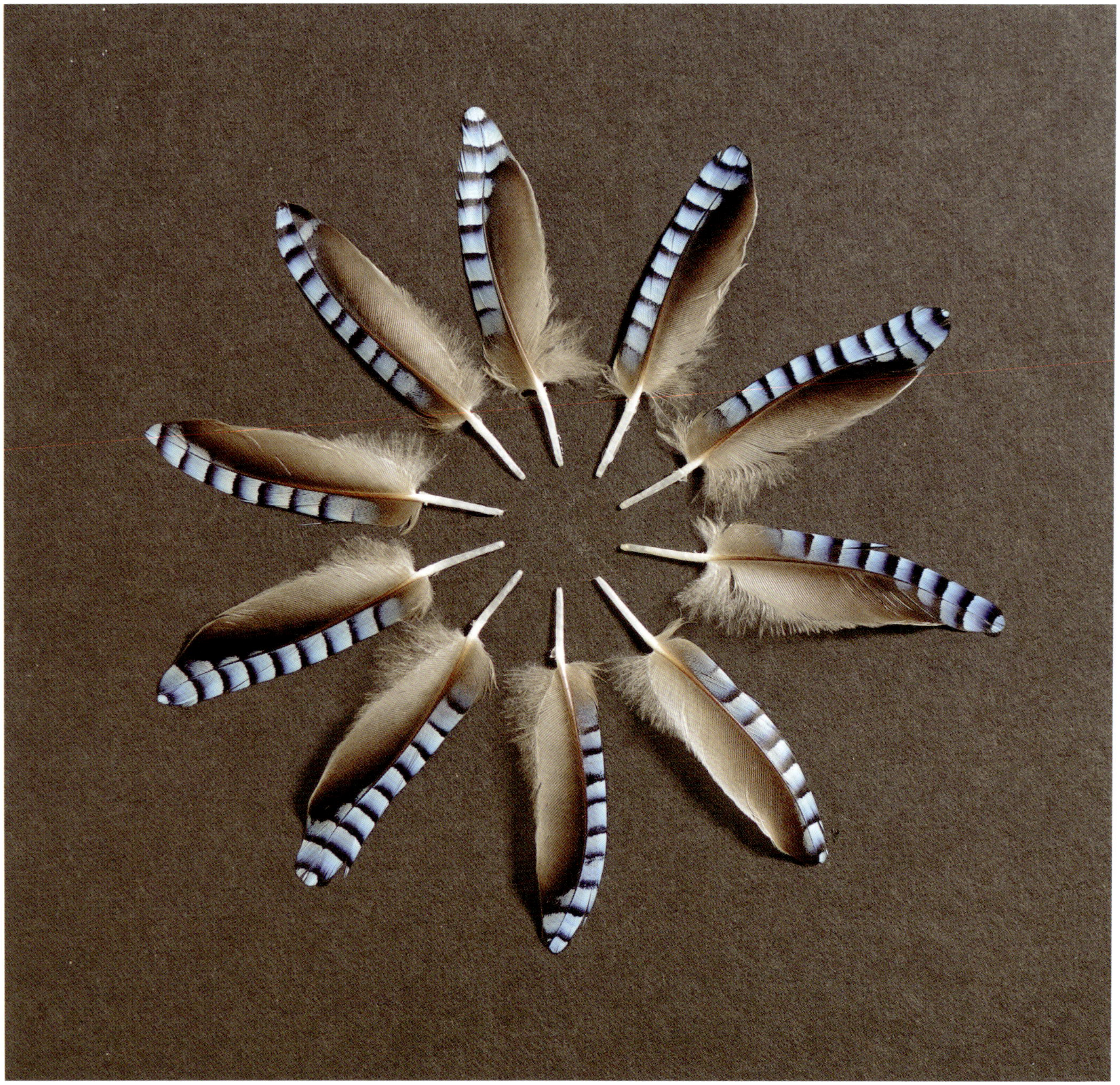

PROPELLER
Eurasian jay feathers
14 x 14 inches

rolling them up while still on the bird and storing them out of the way of dirt. In this way, they can reach lengths of over twenty feet.

These feathers would serve no purpose for the bird in the wild. But in captivity, they serve their purpose to keep these birds alive and multiplying by enlisting human care. In fact, common chickens successfully do the same through producing eggs and meat. According to the UN's Food and Agriculture Organization, there are now 19 billion chickens living on the planet.

A big feather like that of a peacock or an argus pheasant tail takes six months to grow.

When grown, if a feather is pulled out, another grows in. If a feather breaks, the shaft remains there until the next molt, when another growing feather will push it out. When a bird's feathers are all grown, they make up 10 to 20 percent of the bird's bodyweight. How does it take care of all of them?

We use tools like hairbrushes to clean and care for our hair. A bird's tool is its beak (and sometimes its claws), which it uses to take care of its thousands of feathers. With their beaks, birds preen. Without daily preening, feathers would get dirty. They would not be able to repel water and would look ugly and dull. Without preening, feathers' barbs would come unzipped and, eventually, the bird would not be able to fly. Mites would destroy feathers, starting with the down next to the bird's skin. Feathers would clump together in dirty feather dreadlocks. The bird would be cold.

Thinking about the modest amount of time I spend combing, washing, drying, and fussing over my head's small patch of hair makes me appreciative of the effort a bird goes through to keep its feathers in good shape.

Birds spend hours each day taking care of their feathers. With their beaks, they zip back the Velcro vanes, get rid of dirt, and spread oil from glands at the bases of their tails. They remove feather mites with their beaks and with dust baths. Many even condition their feathers with small special feathers that grow and break down into dust on the birds' bodies for this purpose. Birds remove the waxy sheath material from newly grown feathers. It's a lot of work, but it's worth it because clean and functioning feathers keep birds alive.

Flight is the first thing we usually think of when we think about feathers. The Velcro-like barbules join the individual barbs together to make for flat and, therefore, aerodynamic surfaces. Each feather has a special shape that contributes to the specific needs of flight depending on where it is on the wing and what the bird does for a living. The shapes are closely related to the requirements of where the birds live and what their habits are. For instance, curved wingtip feathers of some ground birds help them to take off explosively. You'll find these feather shapes on grouse, quail, and pheasants. Such feathers are shaped more like butter knives than the sharp pointed wingtip feathers of swallows and seagulls, making them more suited for deft movement, sustained swoops and glides.

All feathers and their barbs grow in one direction like fur, making for aerodynamic wings and bodies. If you take your hand

and rub a feather in the direction of its tip, it's like petting a dog in the direction its fur lies. If you rub the fur the other way, however, petting doesn't work, and usually the dog doesn't like it either. Rub a feather the wrong way and the little hooks that hold the feather vane together come apart and it quickly becomes a mess.

All feathers overlap perfectly. The perfect overlap is especially important on the wings. Even the little feathers covering the bigger flight feathers have to fit in exactly the right place. The leading edge feathers on the wing cover those downwind, which are larger and cover the next layer, which in turn covers the next. They are called covert feathers because by covering successive layers, they allow air to pass smoothly over the wing. If even one feather is out of place, say backwards so it catches the air instead of having the air flow smoothly over it, the effect is similar to what happens when you put your hand out of a speeding car window, only far worse.

Each of these feathers is controlled by an individual muscle and tendon; each feather can be manipulated separately according to flight needs. I can't quite imagine doing this if I had feathers—even with my human endowment of a huge brain. At the base of each flight feather, the shaft arises from deep in the bird's wing near nerves. Long flight feathers act as levers enhancing the bird's ability to detect slight changes in air currents and adjust each feather accordingly.

All this said, flight is only one advantage of having feathers. Birds task their feathers with many other jobs.

Feathers repel water. On diving birds, feathers have to be absolutely watertight. Each feather overlaps, forming geometric patterns of diamonds and shingles that trap the downy lower portions of the feathers next to the birds' bodies. You can think of the tips of the feathers as birds' water repellent coats and the downy parts of the feathers as their long underwear and sweaters.

Built into each feather is a microscopic surface structure that repels water using the physics of surface tension. Tiny bumps of keratin protein extend from the feather's surface, touching the water in so many places that it prevents water

LAUNCH
Argus pheasant wing tip feather
11 x 14 inches

All fifty-five species of pheasants run around on the ground and only occasionally explode into the air using their curved, scoop-shaped wing feathers.

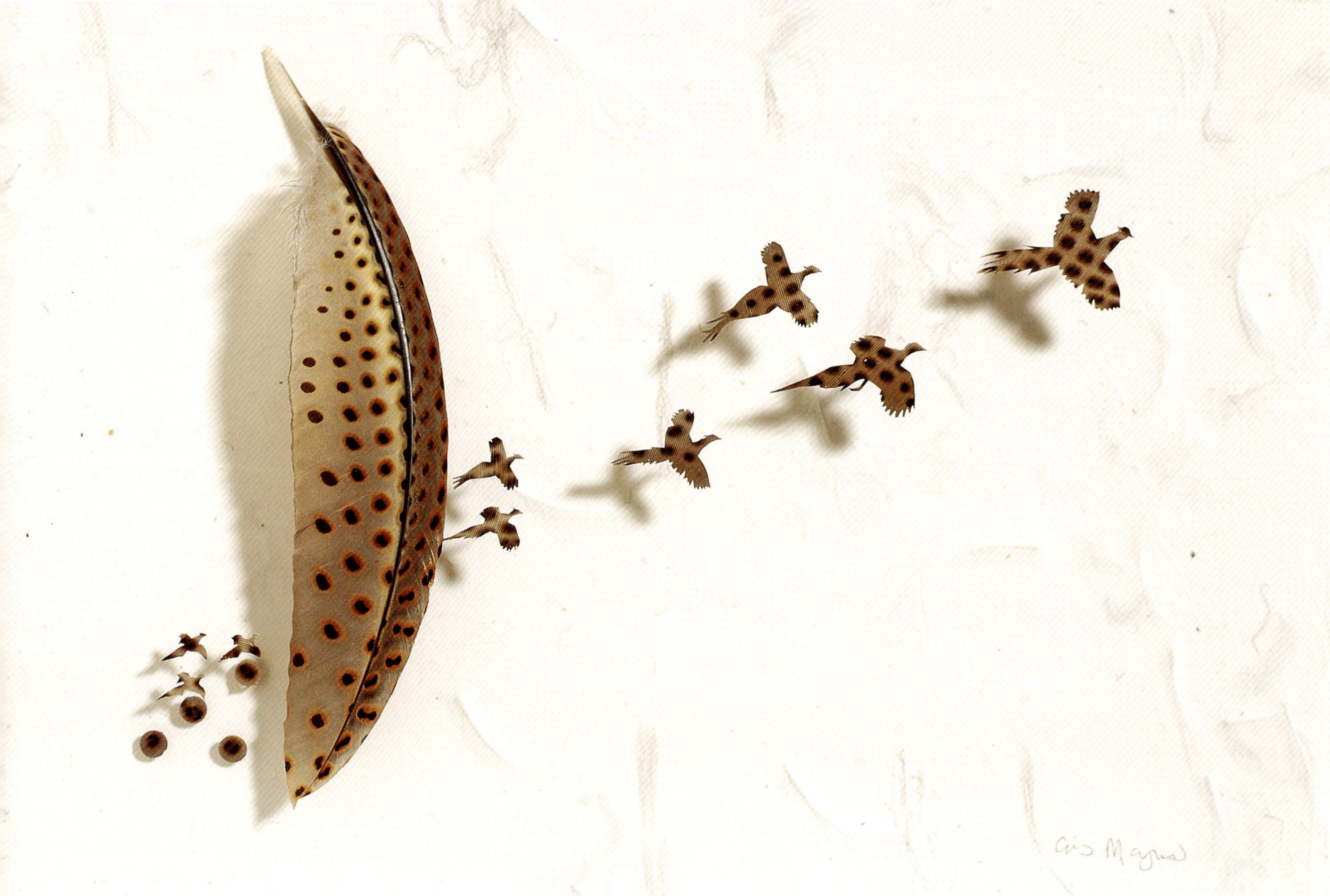

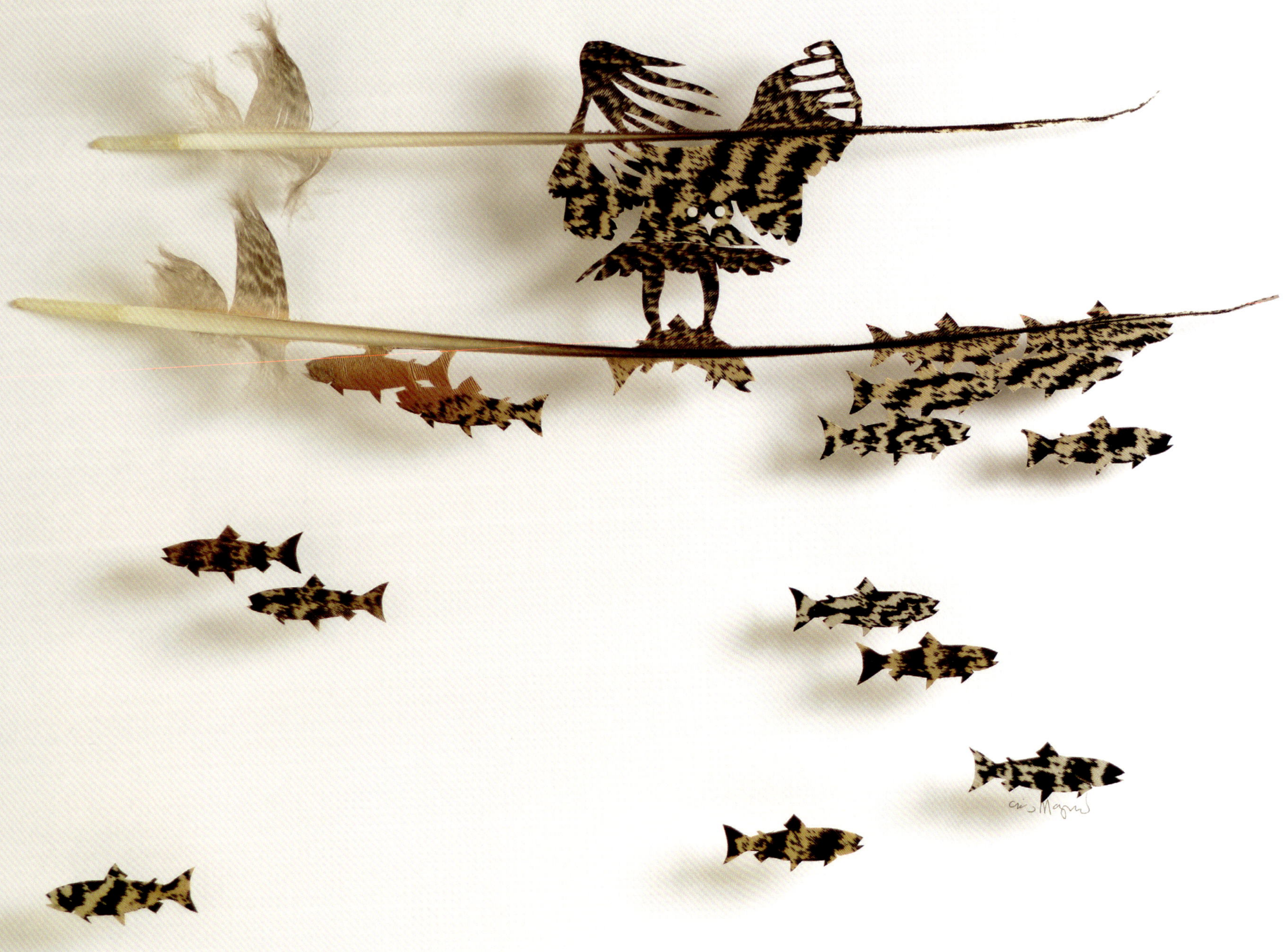

(OPPOSITE)

EAT 9

Peacock wing feathers
15 x 19 inches

Sometimes an osprey will disappear completely underwater when it dives and catches a large fish. When the bird flaps its wings to become airborne again, water is dripping from the feathers. But they are not soaked—just coated with water which drips off and in mid-flight, forcibly shaken off with a shiver or two.

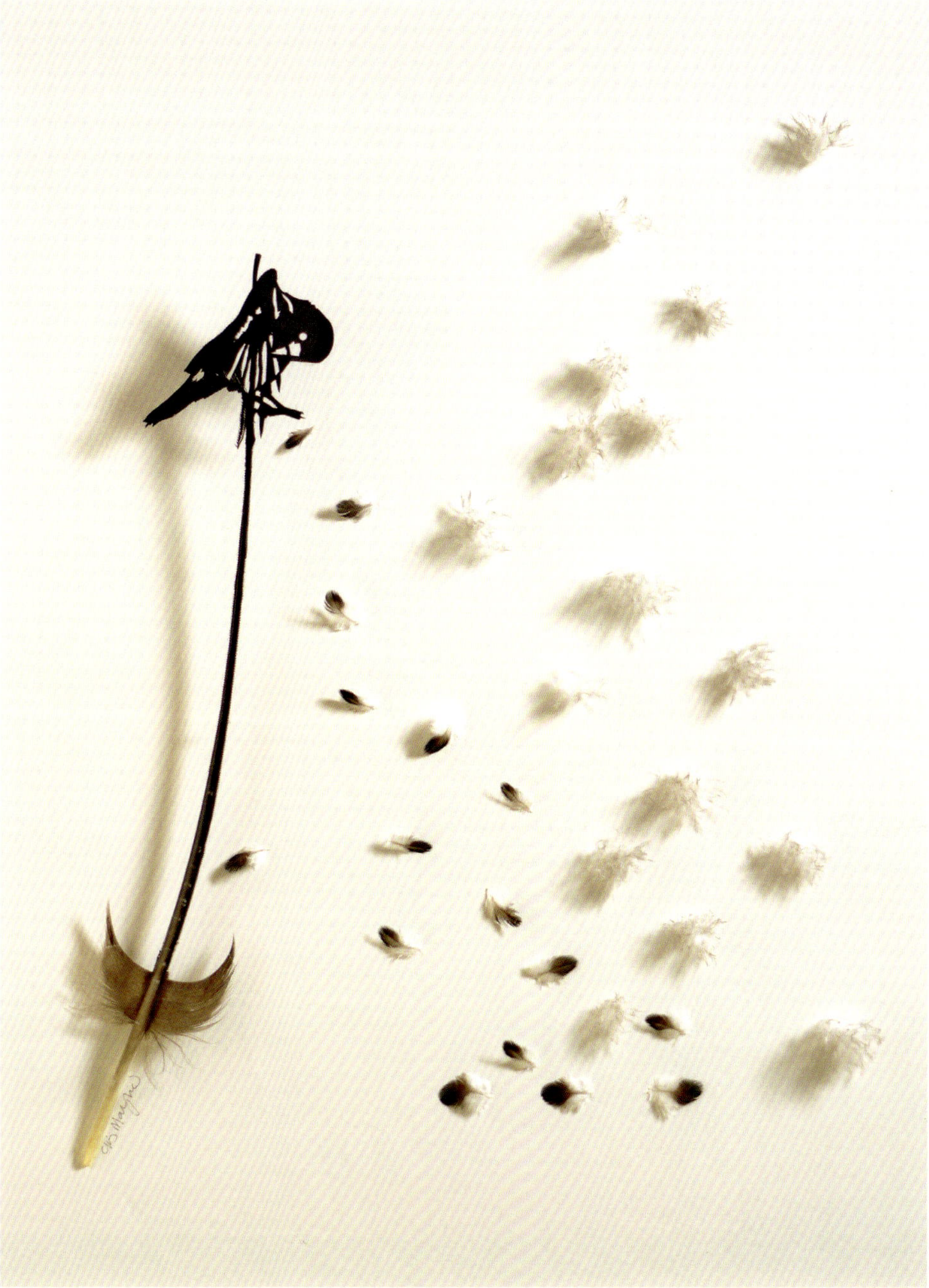

CROW PREEN

Capercaillie grouse feather and parrot body feathers and down
17x14 inches

Why do we call these feathers "down" when they float in the other direction? How about a nice warm "up" jacket, sleeping bag, or pillow?

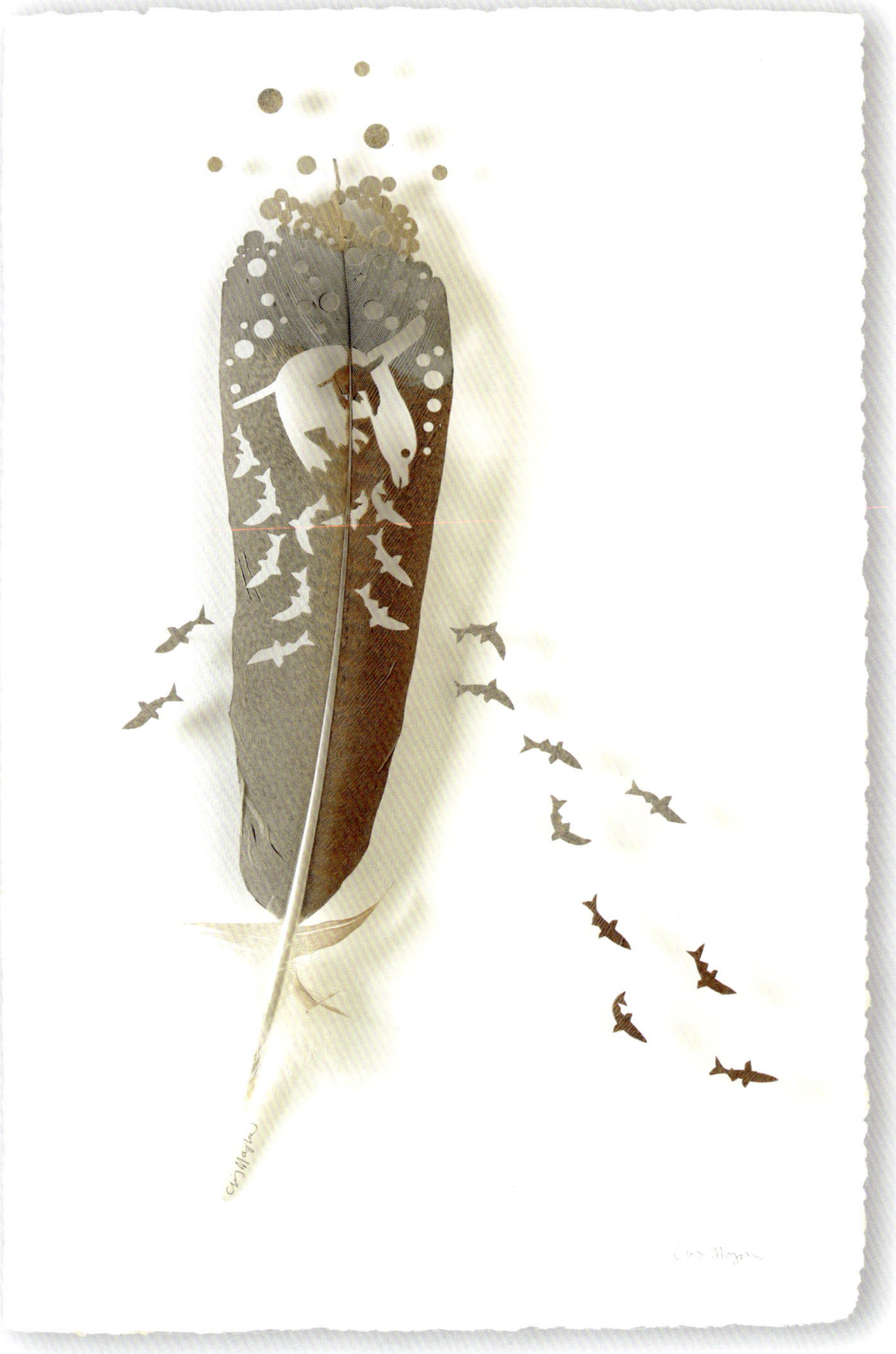

from getting through. The bumpier the feather's surface, the more surface tension it creates. Put a drop of water on a feather and it will bead up, not soak in. Fill a tall glass of water and put a long wing feather in and it will pop right back up to the surface; if you hold it under and observe it, you will see why. A thin film of air adhering to the feather will give its surface a silvery sheen.

I used to think that birds achieved waterproofing by applying a film of oil through their preening, just as I waterproofed my hiking boots with a coat of oil. Not so. Oil conditions feathers and makes them supple, but it does not contribute significantly to waterproofing. It's an interesting footnote that now we have Gore-Tex boots that keep out water, another microscopic method for waterproofing, but still not in league with the elegance and effectiveness of feathers.

> Being waterproof and thereby having dry down keeps a bird warm.

Since only the waterproofing tips are exposed, underneath lies air-trapping down. The shafts on most body feathers curve both front to back and side to side. On water birds, like ducks, body feather shafts may curl more than halfway around, over 180 degrees: it takes two to make a complete circle. The curve allows a larger air space between the body and the elements. As with the insulation in your house or the layers or "loft" of your clothing, the thicker it is, the warmer you are. Birds can control how thick the air layer is. They can puff out, trapping warming air. And through muscle-feather control, they can flatten the flexible shafts, pushing them next to their body.

I often take binoculars on my walks around the lake near where I live. On a cold day in November, I watched some diving ducks disappear underwater and a minute later pop up like corks. As I observed them, I noticed that the split second before they dove, their bodies seemed to shrink significantly. On the surface, this phenomenon seems the opposite of what I do right before I dive into the water. I want to puff up my lungs with a breath of air to extend my oxygen supply. But I saw that the duck's preparation had nothing to do with breathing: it flattened its curved body feathers to squeeze out the air that would otherwise make the bird buoyant and pull it up to the surface.

(OPPOSITE AND RIGHT)

PENGUIN SWIM 1 AND SWIM 2

Turkey feathers, lilac turkey breed
18 x 12 inches

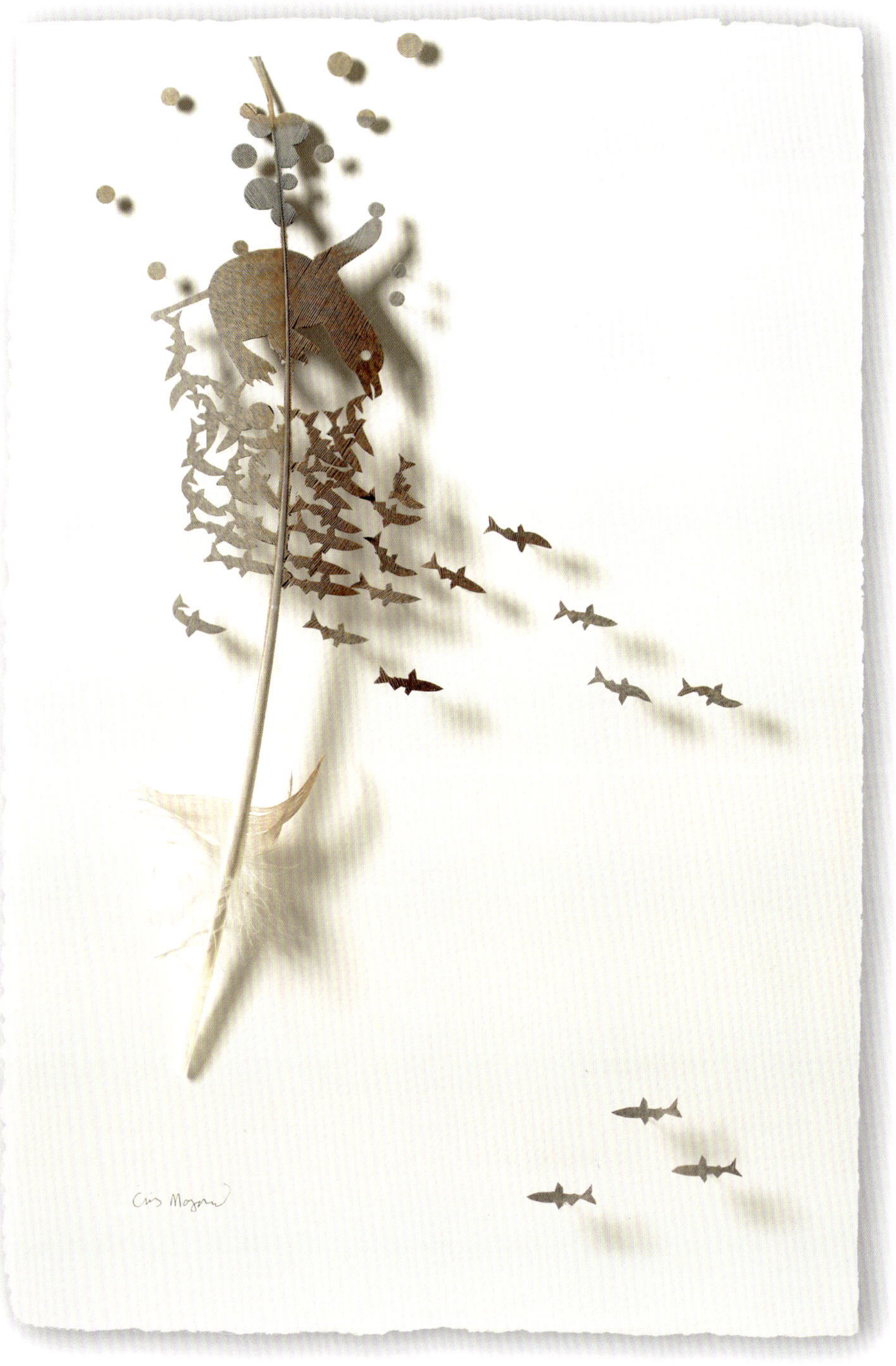

This same cold day, I saw songbirds perched on low branches looking especially cozy as they fluffed their feathers. As these birds erected their feather shafts away from their bodies, they were allowing the natural curl of their feathers to spring out, trapping more air and keeping them warmer. Just looking at them had me wishing I could do that with my clothes.

On a hot summer day, I wear minimal clothes more for social propriety than for warmth. Public nudity hereabouts is not an option. But regardless of heat, I may still opt for more body coverage with clothing to protect me from getting sunburned. Birds, though, need not be troubled by this type of social nicety. The permanence of the feathers does the deciding—a good thing. What would a featherless bird look like after a busy day in the sun while navigating thick brush? It would be sunburned, bruised, cut up, and scraped.

So indeed, protection from scratches may be closer to the primary use of feathers for some birds. Because rails and bitterns spend their days squeezing through cattails and rushes, their feathers wear quickly. These birds molt their feathers several times a year. Though flight is more important in the daily life of a honey buzzard, its feathers enable it to eat bee, wasp, and hornet larvae by shielding its body from stings. Feathers are a bird's armor, insulating it from environmental challenges.

Yet feathers also do the opposite: they help the birds connect more intimately with their environment. Each body feather functions as a sense-augmenting lever by transmitting pressures and sensations down the shaft to sensitive nerves near its base in the bird's skin.

Even though one of the benefits of being a bird is the ability to fly away from danger, birds' feather coloration and pattern give them the alternative capacity of being able to hide undetected. As an artist, I find patterns in bird feathers intriguing. Feathers off the bird are striking, especially those with patterns. But when birds wear their feathers, they blend into their habitat—be it woods or brush, sandy desert or shore, marsh, or grassland—and seem to disappear, camouflaged.

> Feathers help birds in ways that might not come immediately to mind. Like human clothes, feathers provide protection.

OOPS
Turkey feathers
13 x 16 inches

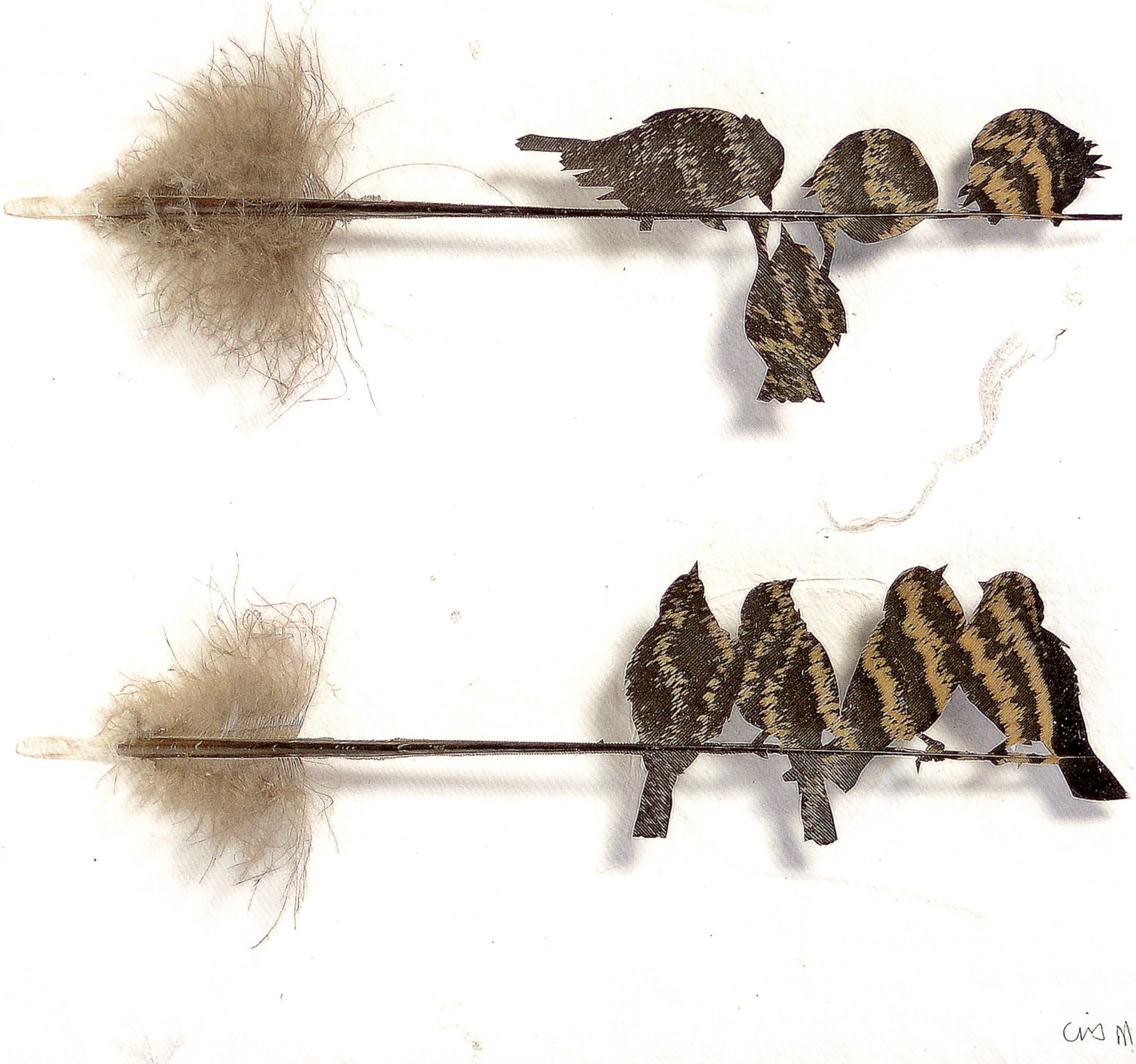
Chris Maynard

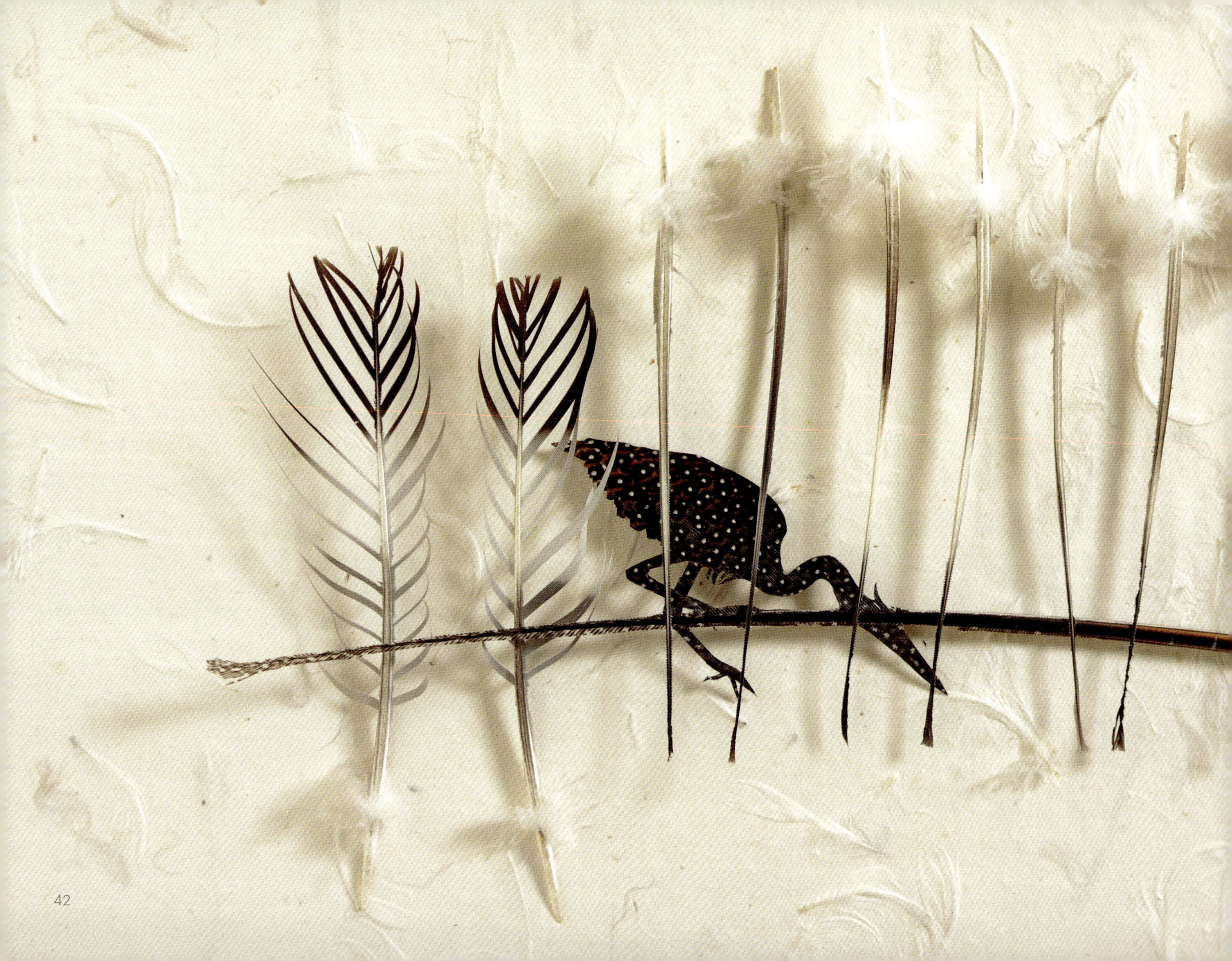

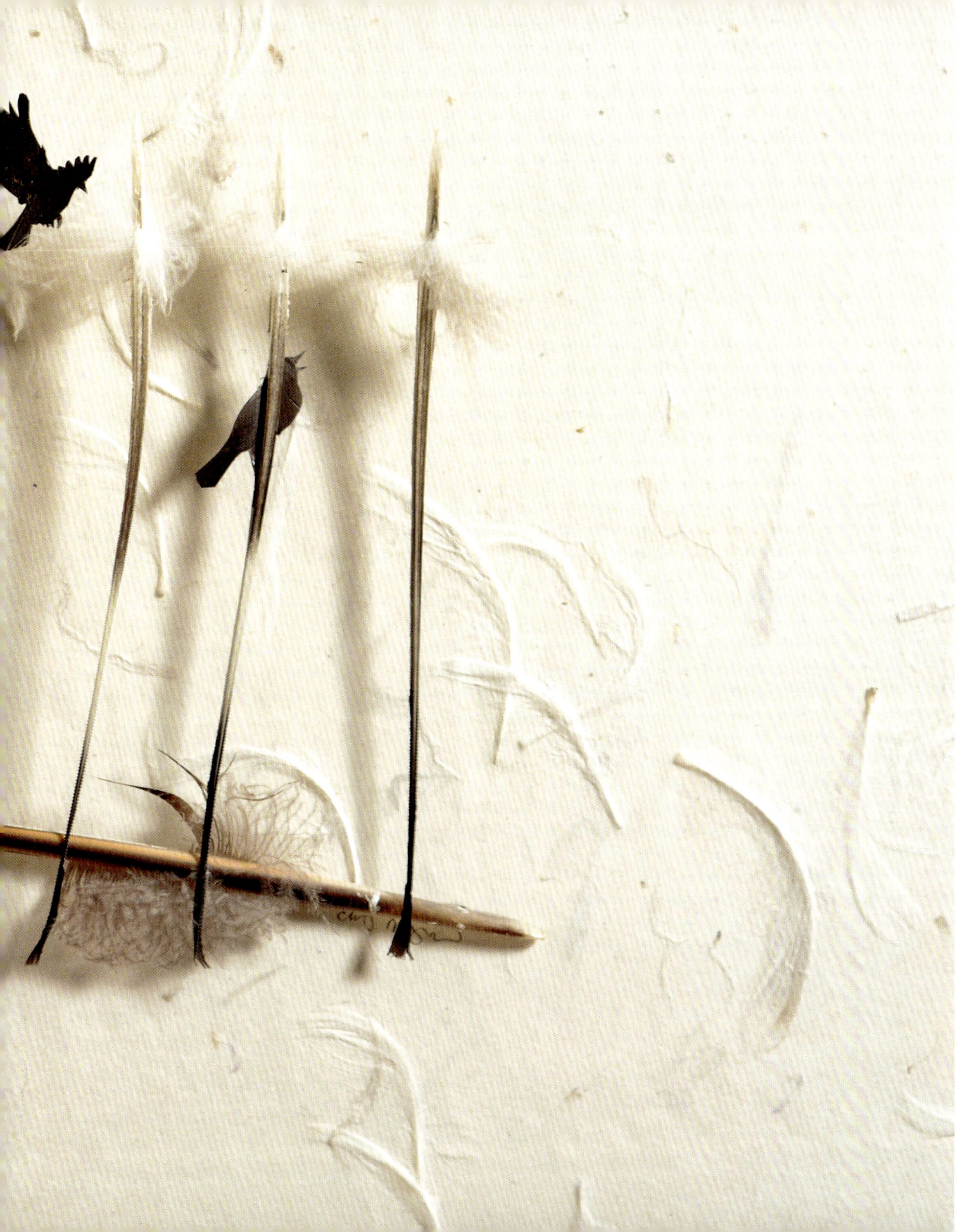

HERON LAND
Argus pheasant feather
and pigeon feathers
14 x 21 inches

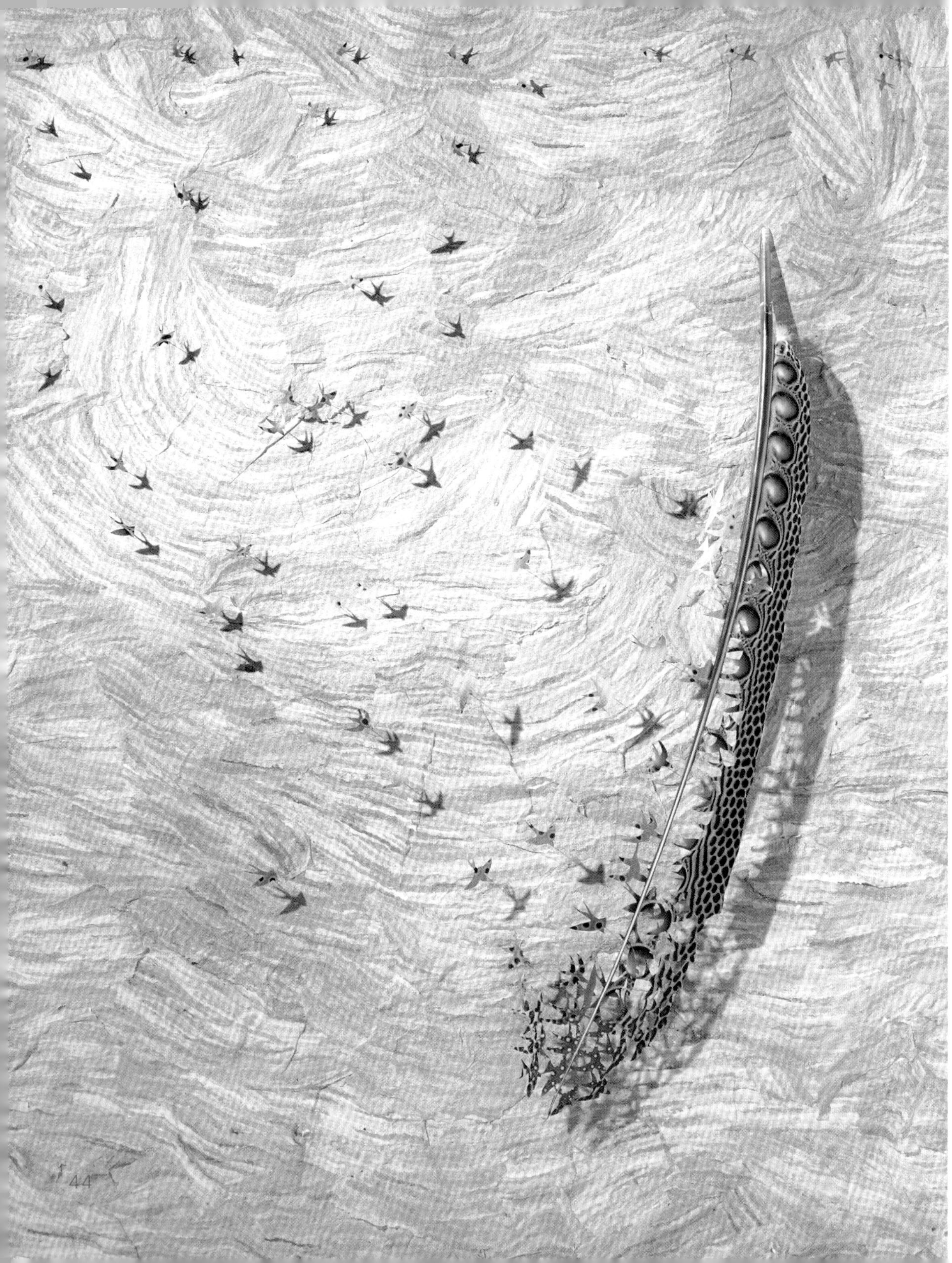

SHELTER 2

Argus pheasant feather
32 x 26

A bird's feathers are its house. They are roof tiles to shed rain, walls to keep out the wind, and down insulation for warmth. The background here is paper made by wasps. A large nest (minus the wasps) was carefully unpeeled layer by layer, then flattened and pasted onto watercolor paper.

Recently, I sat high in the Uinta Mountains of Utah on a cold clear November day and watched a ruffed grouse slowly forage around me, an arm's length away. At first, I heard a rustle under the juniper brush and aspen trees and stared for five minutes at the spot ten feet away. Finally, after another small rustle and movement, its shape emerged. If clothing designers took note of what birds wear, I would buy a jacket and pants in ruffed grouse.

I found such an extraordinary abundance and diversity of coloration and pattern in the individual feathers of the great argus pheasant that I was able to find the entire alphabet plus all the numbers embedded in my assortment of argus pheasant feathers (*image page 46-47*). This bird's feathers sport so many patterns that I found not only one but many As, Bs, and Cs. When these feathers are on the bird, in the woods in its native Viet Nam and Laos, it is very difficult to spot. I have never visited the native habitat of the great argus pheasant, but I got a taste for how well camouflaged it can be while observing the grouse in Utah.

Feathers on birds' bodies overlap each other so just their tips are visible. These tips make patterns. Depending on the tips' shapes, they are seen on the bird as repeating diamond, triangle, or sometimes rectangle shapes. Patterned feathers' tips add greatly to the total effect. A single patterned feather is a puzzle piece that fits together seamlessly into the surface of a bird's body like shingles on a roof or a tessellation illusion by the Dutch artist M.C. Escher (*image:resurgence page 64*).

Trying to improve my feather alphabet by using more colorful feathers than the brown and black melanin colors of the argus pheasant, I failed to find much. Why? Colors in feathers make birds stand out. They are for display. Patterns, on the other hand, hide a bird. Therefore, colors with patterns are rarely found in feathers. Colors let birds advertise themselves to mates and potential rivals. When predators come close, birds with patterns tend to hide, while birds with bright colors tend to fly away.

After my failed attempt at the color alphabet, I began looking for more information on feather color and came across the research of Dr. Richard Prum of Yale University. Reading his papers, I began to learn about how the surface texture of some feathers makes them so colorful. Here is what I found.

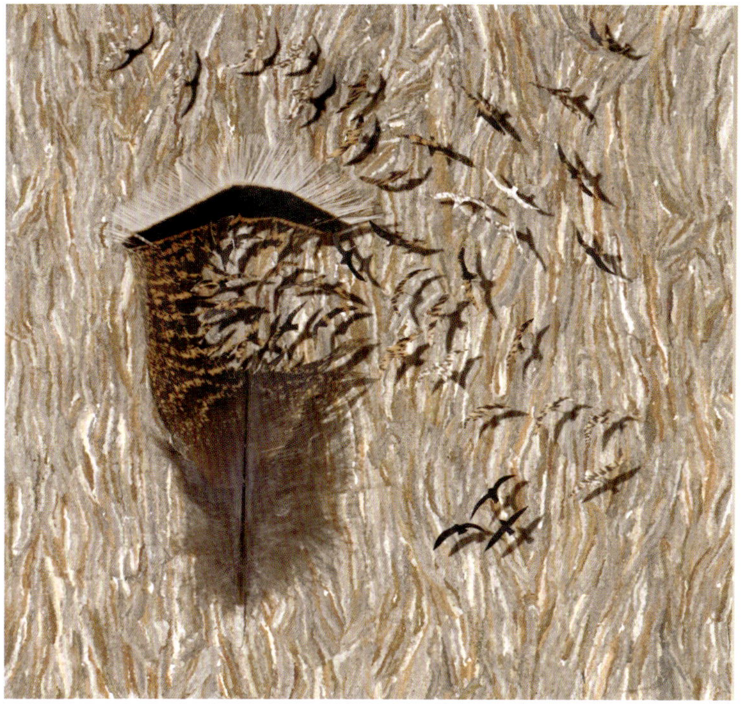

SHELTER 1
Turkey feather
14 x 11 inches

During a snowstorm, I discovered a large, almost abandoned wasp nest in a barn. Bringing it in by the warm woodstove, I carefully took it apart. A few sleepy wasps crawled out. The wasps did not try to bite or sting anyone; they just wanted to hibernate for the winter. This paper was not hand-made, but was mouth-made by wasps chewing wood, mixing it with their saliva, and flying back to their home to empty the mix next to their last spittle to slowly create their nest.

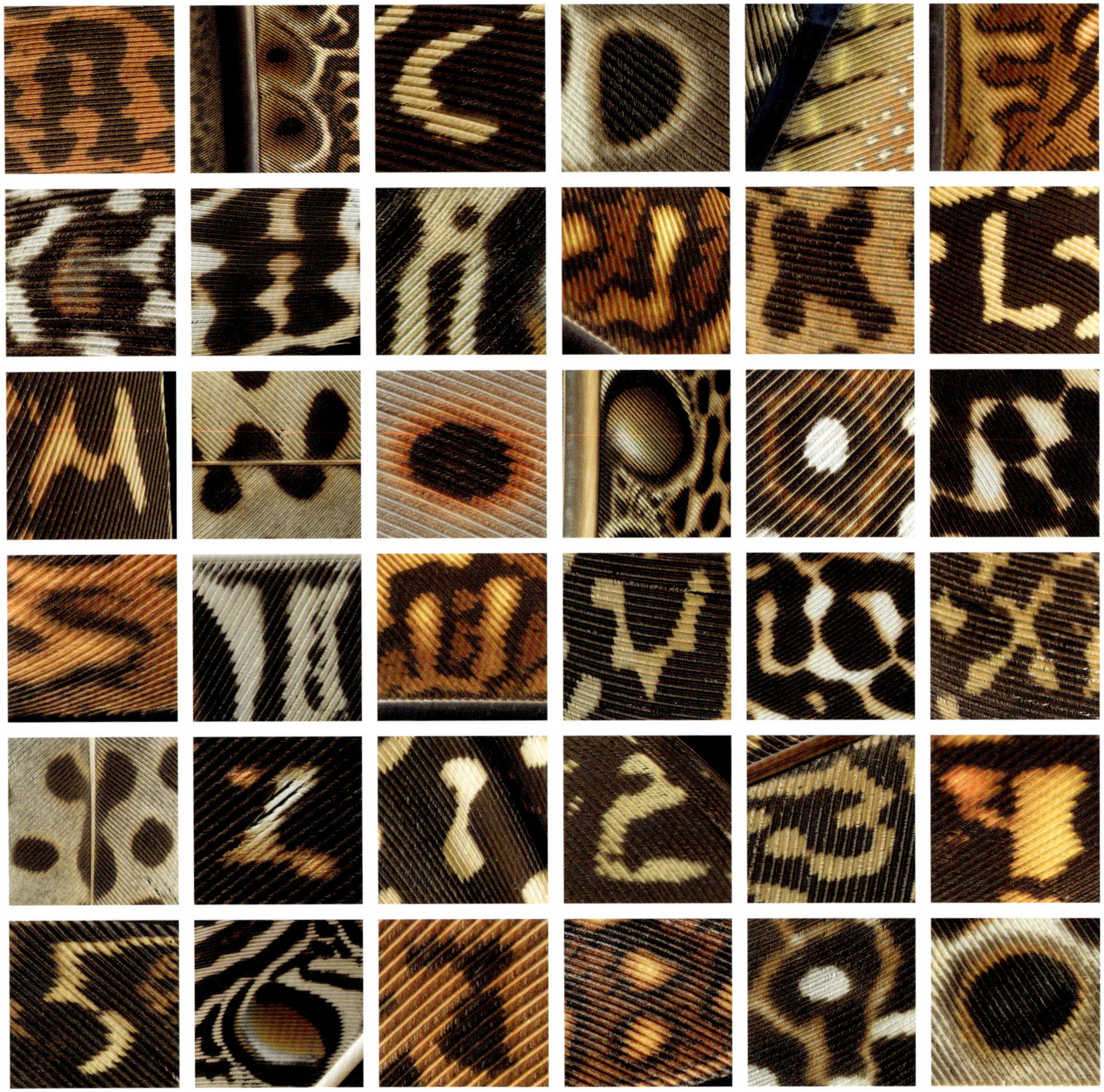

ARGUS ALPHABET, A POSTER

Great argus pheasant
male, female, and juvenile feather patterns
25 x 18 inches

Recognizing each letter and number takes a bit of eye-work. Discerning the live bird is also difficult as it uses these same patterns to blend into its forest background.

Individually, the wing patterns in the argus pheasant are striking, containing the entire American English alphabet and, if one looks closely at enough feathers, Chinese characters and Cyrillic and Arabic scripts would likely be found as well as Devangari, Kana, Hangeul, and many other scripts of the world. The birds don't make these symbols for us; we just use our imaginations.

AMAZON CIRCLE

Amazone parrot wing feathers
16 x 12 inches

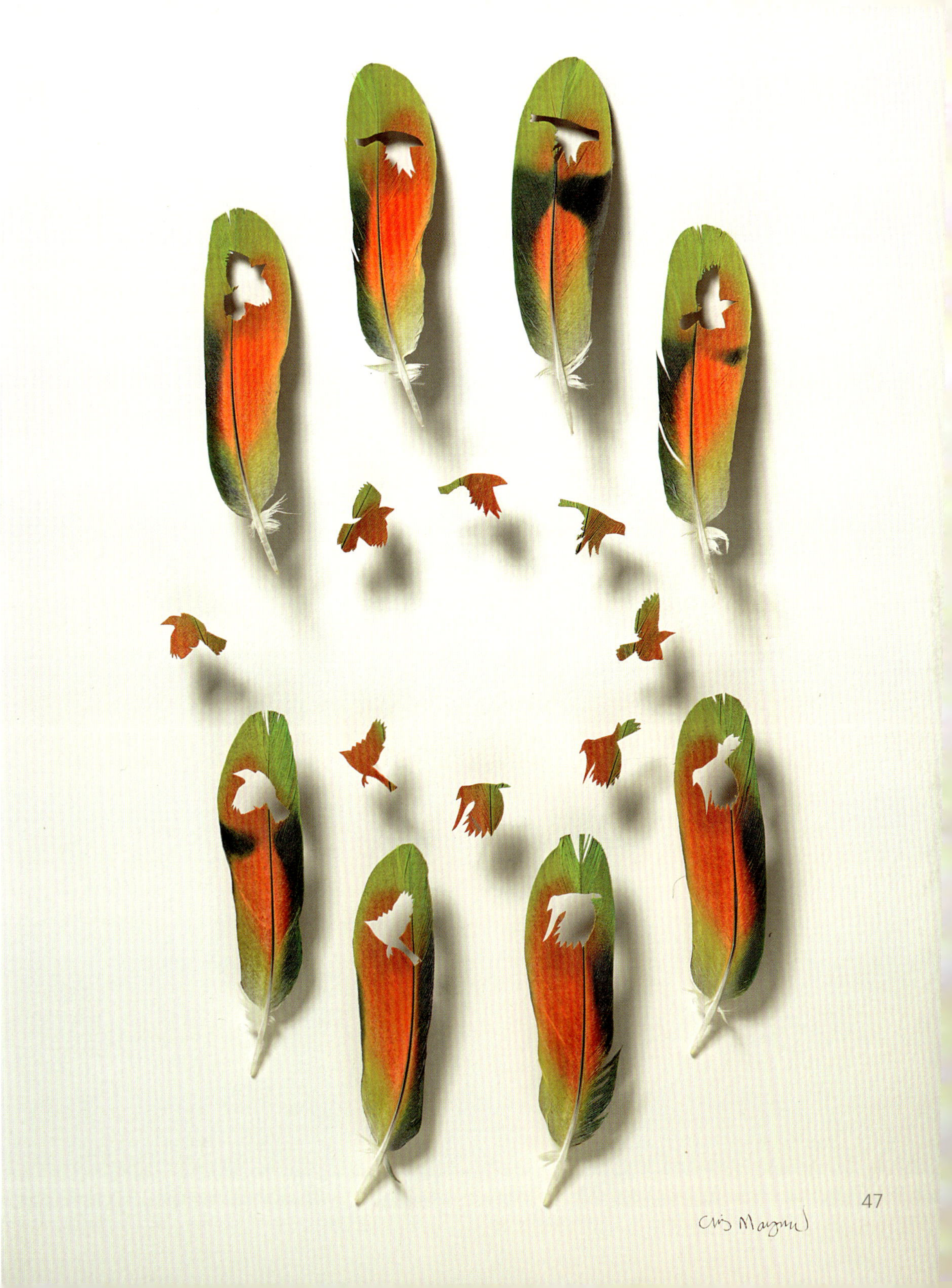

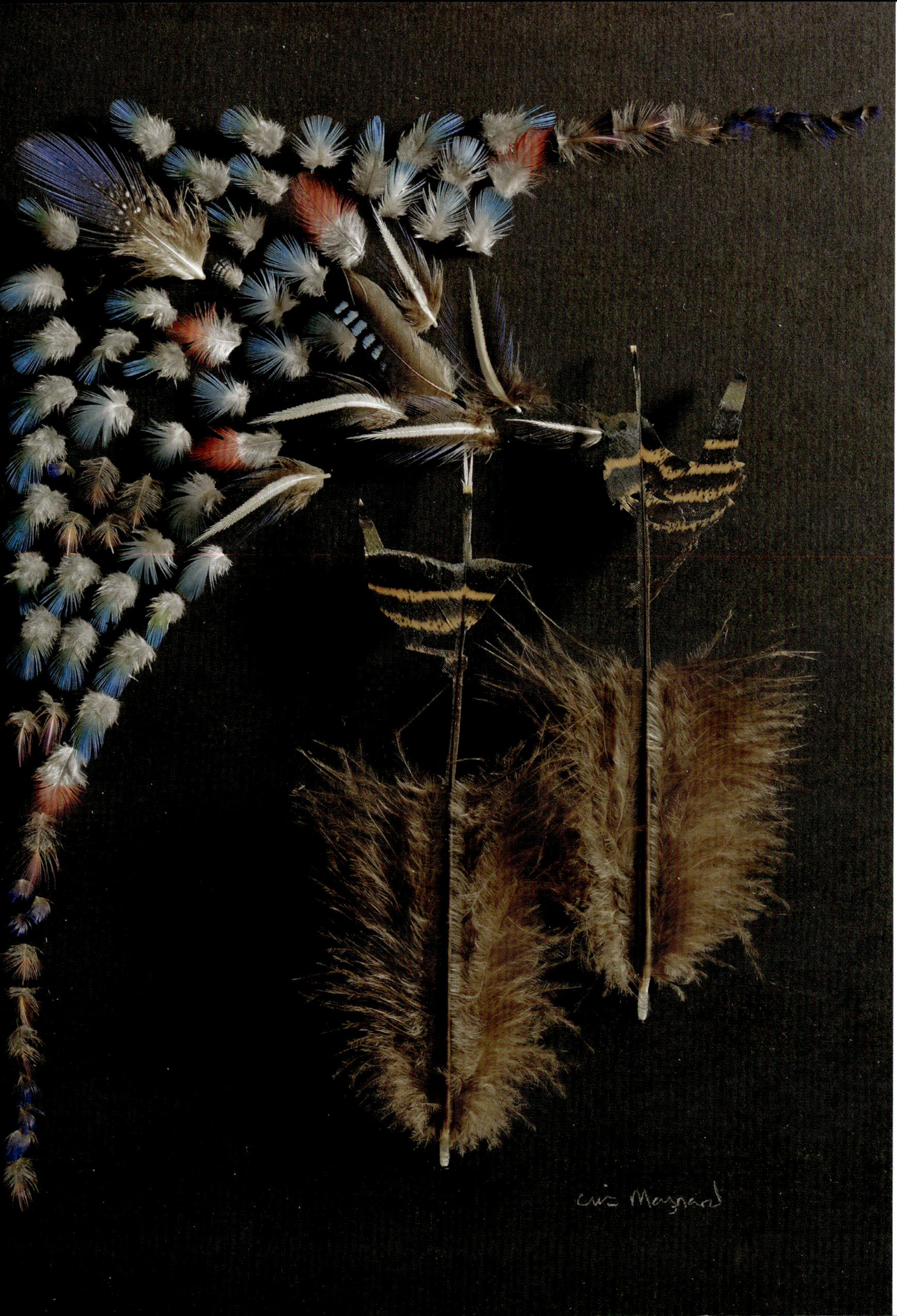

All colors, including those in feathers, are made in one of two ways, either by chemicals or from microscopic structures formed on the feather's surface. Prum's team, comprised of an ornithologist, electrical engineer, and physicist, is unfolding a whole world of how structural colors look and work. The team's work has revealed how we see structural colors in feathers as the metallic and iridescent green, gold, blue, and purple of a peacock's feathers' eye, in the metallic reflective colors in starling feathers, and the blue of a blue jay or macaw parrot. It turns out that we see metallic colors as a result of light reflecting from a variety of nano-scale surface grooves that reflect light of certain wavelengths from one or more directions. A peacock feather appears shiny from only one angle because the shape of the grooves in their surface structure reflect light from only that angle. The pattern

---

NIGHTINGALE
Two turkey feathers and small feathers from a parakeet, vulturine Guinea fowl, cotinga, Eurasian jay, and a lilac-breasted roller
14 x 11 inches

Made on December 21st, the darkest day in the northern hemisphere.

When predators come close, birds with patterns tend to hide while birds with bright colors tend to fly away.

and size of the nanostructures determine what color you see. Many blue colors such as those found in macaw and bluebird feathers also result from structures on the feathers' surface, but they do not appear metallic at all. One sees the same blue color from many angles. This is the result of a spongy surface microstructure full of holes. Remarkably, and in a way not yet completely understood, these holes capture and distribute light evenly through the feather's spongy surface and only let blue light return through their openings to our eyes. Feathers share this class of structural colors with fish, plants, and insects. We will likely someday find these and future discoveries industrially massaged into our lives in ways that include printing, computer screens, fabrics, and paints.

As a visual artist, I am often profoundly aware of the ways colors enhance the world's beauty. Humans have three kinds of cone receptors in their eyes that allow us to see primary colors. Birds have four. They can see ultraviolet and violet colors we cannot. Their eyes are different from ours in other ways too: they can see subtle differences in primary colors. To them, primary colors appear sharper, more vivid. It would seem that feathers have adapted to use the full range of birds' visual abilities. In other words, to a bird, feathers are even more colorful. A black feather may appear to a bird as a shimmering wonder of violets and blues unknown to us. Imagine my disappointment when I learned this. I want to see like a bird. Instead, I have to settle for using my imagination.

I recently interrupted a rather long spell of reading about feather coloration to go to the grocery store, still imagining on my trip how female birds seek out perky, bright, saturated colors when selecting their mates and how brighter colors indicate bird vitality and health. At the store, I realized I was looking at shoppers from a slightly different perspective, noticing more about their choices of colors in their clothing. Perhaps we are not that much different than birds in using color to attract attention, although we are fortunate that we can change our colors as easily as we change our clothes. Next to the checkout line, the store displayed the ubiquitous women's magazines. The cover of *People Magazine* showed a photo of Christie Brinkley that read "Christie at 60! How she does it!" She was pictured in a bright blue bathing suit, looked stunningly beautiful, and didn't seem a day over twenty-five years old. It occurred to me that in addition to changing clothes, we now have another way to make ourselves look how we want: Photoshop.

People Magazine pictured the blue in Christie Brinkley's bathing suit not by using structural colors in their printing process but by using chemicals. Birds also employ chemicals to make most of the reds, oranges, and yellows in their feathers. Most of these are metabolized by the birds from carotenoids, a pigment found in what they eat. Birds get these carotenoids mostly from plants and other animals like shrimp, which makes a salmon's flesh red and scarlet ibis and flamingo feathers orange.

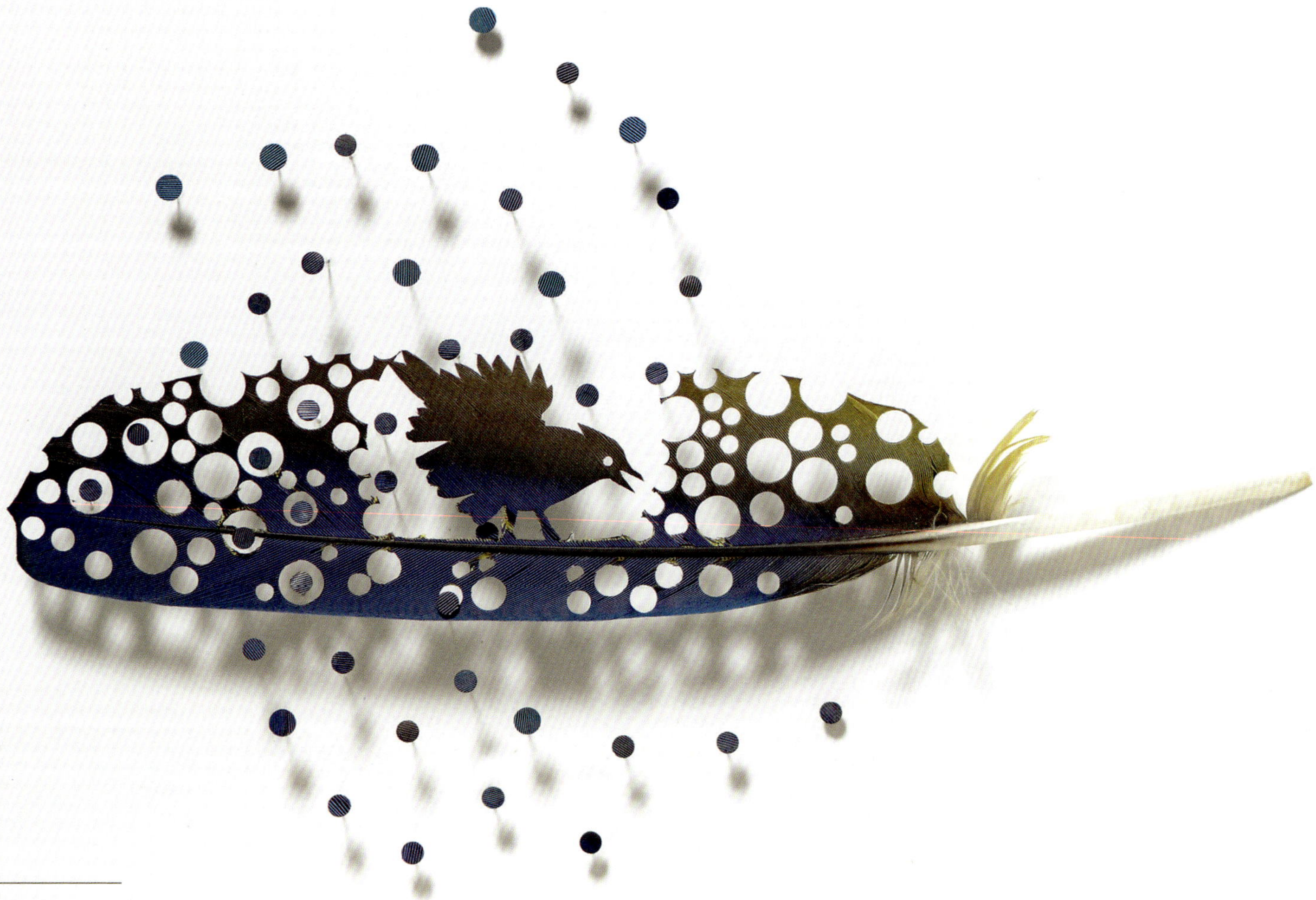

JAY SUNBATHER

Blue and gold macaw feather
11 x 14 inches

A jay needs its bath, just like you and me. But bathing is just a start. After its bath, the bird flies to a safe perch to preen for an hour, finishing the work of keeping its feathers in impeccable shape.

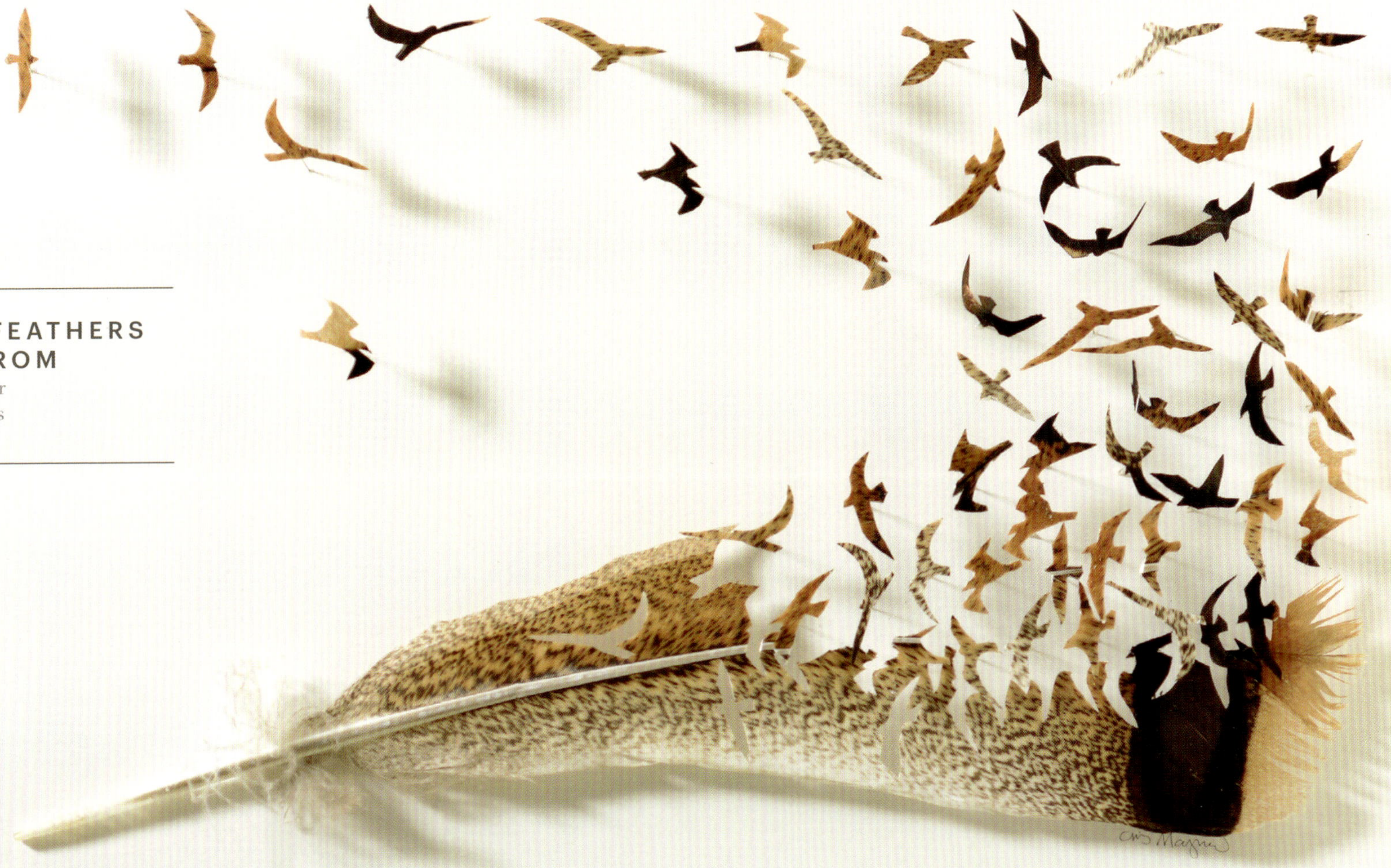

WHERE FEATHERS COME FROM
Turkey feather
20 x 25 inches

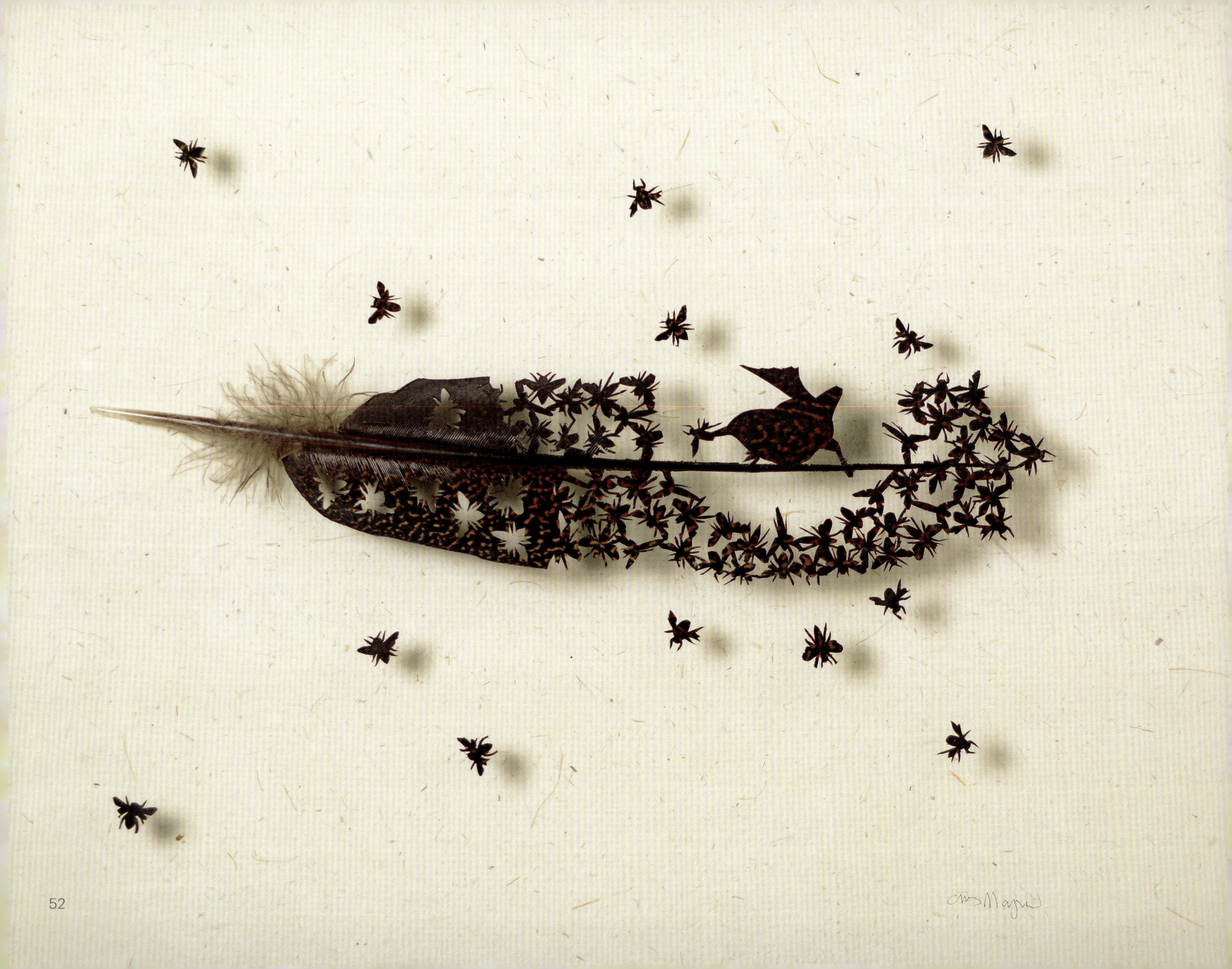

Colors make birds strikingly beautiful and are relatively scarce when compared to the browns, grays, and blacks of most feathers. These common feather colors come from melanin, the same pigment in our hair and freckles. Besides coloring feathers, melanin makes them stronger. Gulls and many birds' white wing feathers have black tips. It's as if they are reinforced there, in the places that get the most wear.

I heard a rumor that the red and green colored feathers of the turaco family of birds are water soluble. I doubted the accuracy of this assertion. I thought, "In the rain, wouldn't the colors wash off of this African forest bird?" I went to the kitchen, rummaged through the cupboard for a clear glass, filled it with water, and placed in it a beautifully red-colored turaco flight feather (the color comes from a copper-based pigment in the fruit the bird eats) to soak overnight. I had to massage the feather a bit to break its surface tension and let water soak in. The next morning, the water was still clear with no sign of any color having washed out of the feather. Then I tried adding a bit of dishwasher detergent to the water,

(OPPOSITE)

EAT 1
Female argus pheasant feather
12 x 17 inches

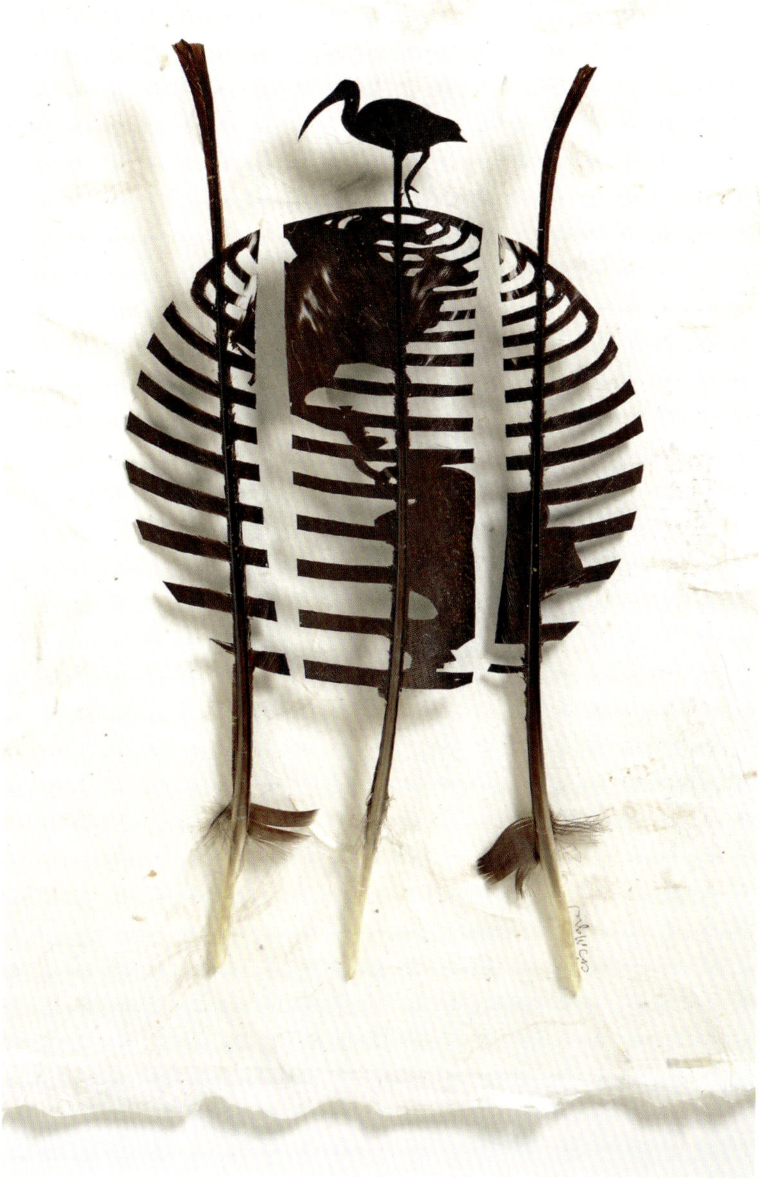

IBIS WORLD
Capercaillie grouse tail feathers
20 x 16 inches

With air travel, distances have shrunk for us and the earth seems much smaller. Birds, such as terns and godwits, migrate across the globe twice a year every year and have done so for eons. How big is the world to them?

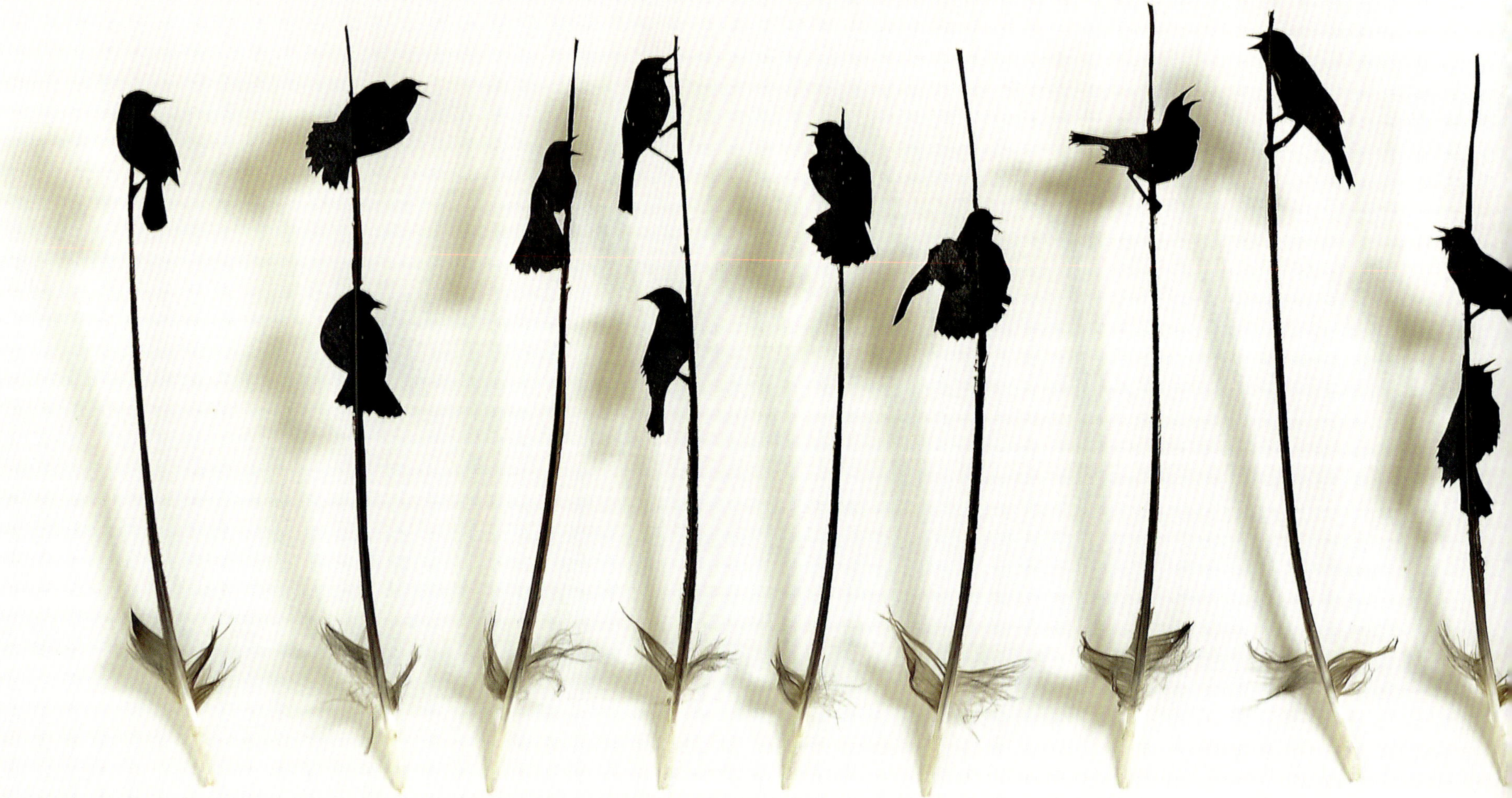

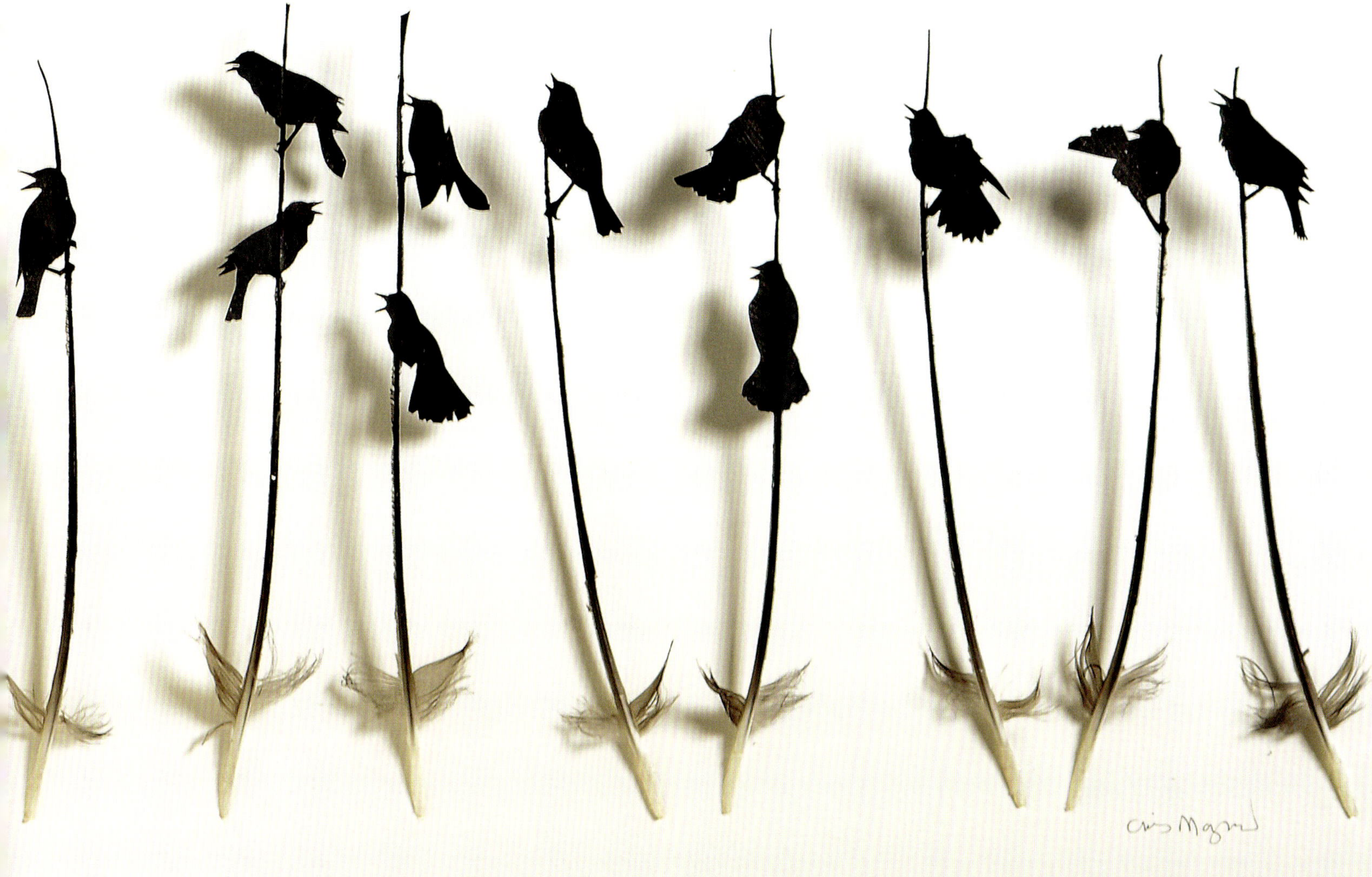

and very soon, a red color began bleeding out into the soapy water. In a minute, the single finger-length feather made the entire glass look like tasty deep red cherry juice. The feather, on the other hand, came out light pink, most of its color washed out.

Some colors fade over time. In addition, feathers wear through exposure to wind, sunlight, and from rubbing against brush and branches. To grow new, fully functioning feathers, most birds shed their feathers once a year; some do it twice, some less. This is called molting. As feathers are shed, new ones grow back in their place. The molt is a hugely stressful time for the birds because it takes a lot of energy to grow back 10 to 20 percent of their bodyweight in feathers. Their molting season is during a time of abundant food to replenish themselves, after spring and summer mating and after giving their chicks a good start.

Birds shed their major flight feathers in pairs, one from each side in an orderly progression as they drop off the bird. A bird would fly lopsided if it molted a major feather from one wing and not the other at the same time. It would be like trying to swim with one hand balled into a fist and the other hand with fingers spread out. If you observe crows or other large birds flying during molting season, some will be missing fully grown flight feathers, one on each wing.

The sequence of the feathers molted varies between species. Often, the first flight feathers to shed are in the middle of the wing and the molt progresses moves both outward toward the wing tip and inward toward the bird's body. On the tail, the two central tail feathers often molt first. The new feathers grow enough to help a bird fly, and only then does the next pair shed. After the last feathers are shed and new ones grown in, the bird plumage is ready for migration, winter, or mating season.

A shed feather is a gift from a bird that may still be living. Take delight in imagining what adventures it has experienced. Perhaps it migrated from another continent, helped to hide its owner from a fox, or helped attract a mate.

(PRECEDING PAGE)

BLACKBIRDS 3
European crow feathers
14 x 32 inches

The male redwing blackbirds roam in large groups during the winter. But come spring, they vigorously defend their small territories from each other.

PLUME
Turkey feather & various tiny parrot feathers
11 x 15 inches

Words are made up of the tiniest sounds—vowels and consonants—arranged together to make a whole meaning. To write, with a pencil, a ballpoint pen, or a feather pen, is to try to capture a reflection of the real thing we are trying to describe. But the real thing always escapes capture.

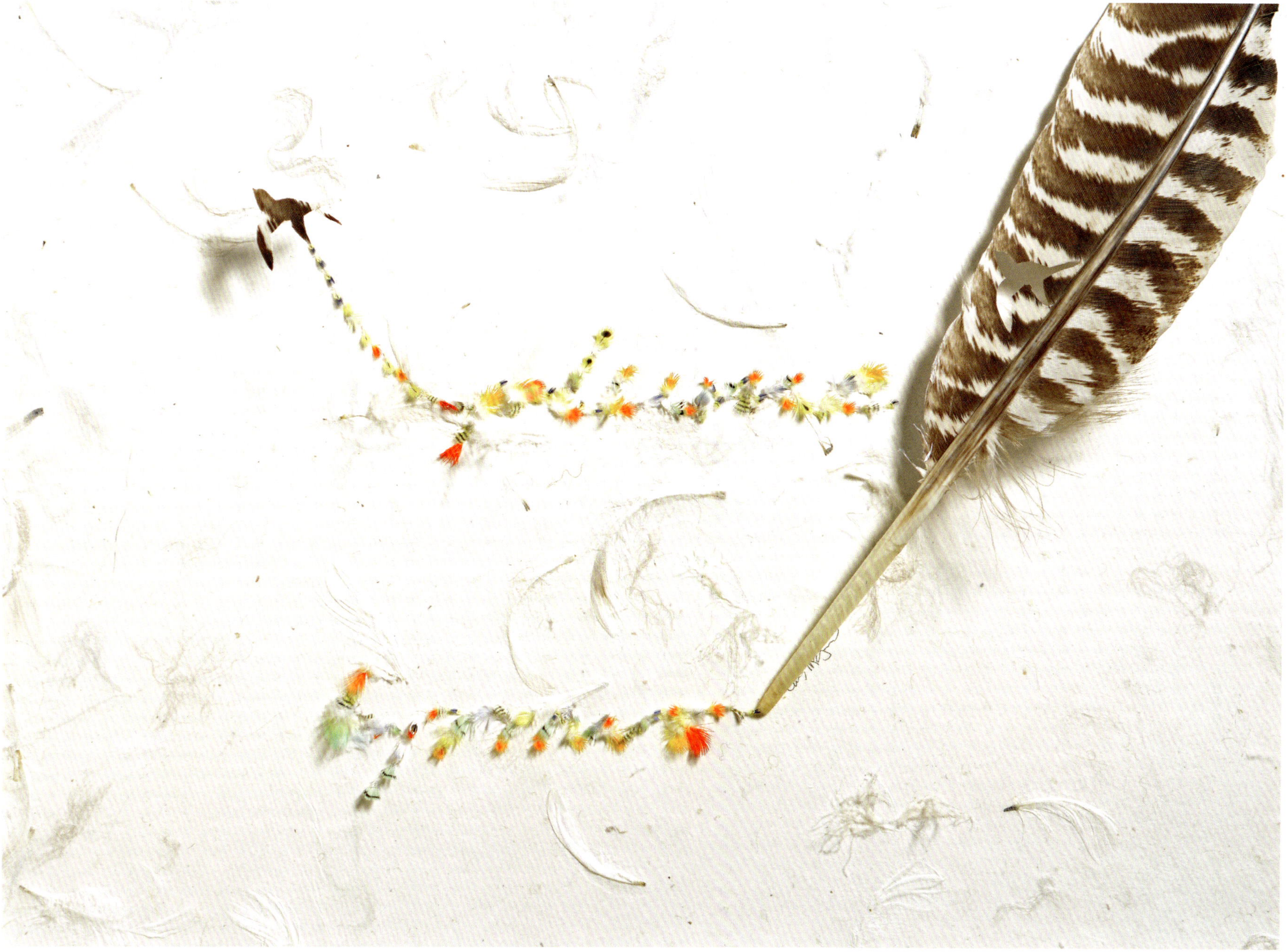

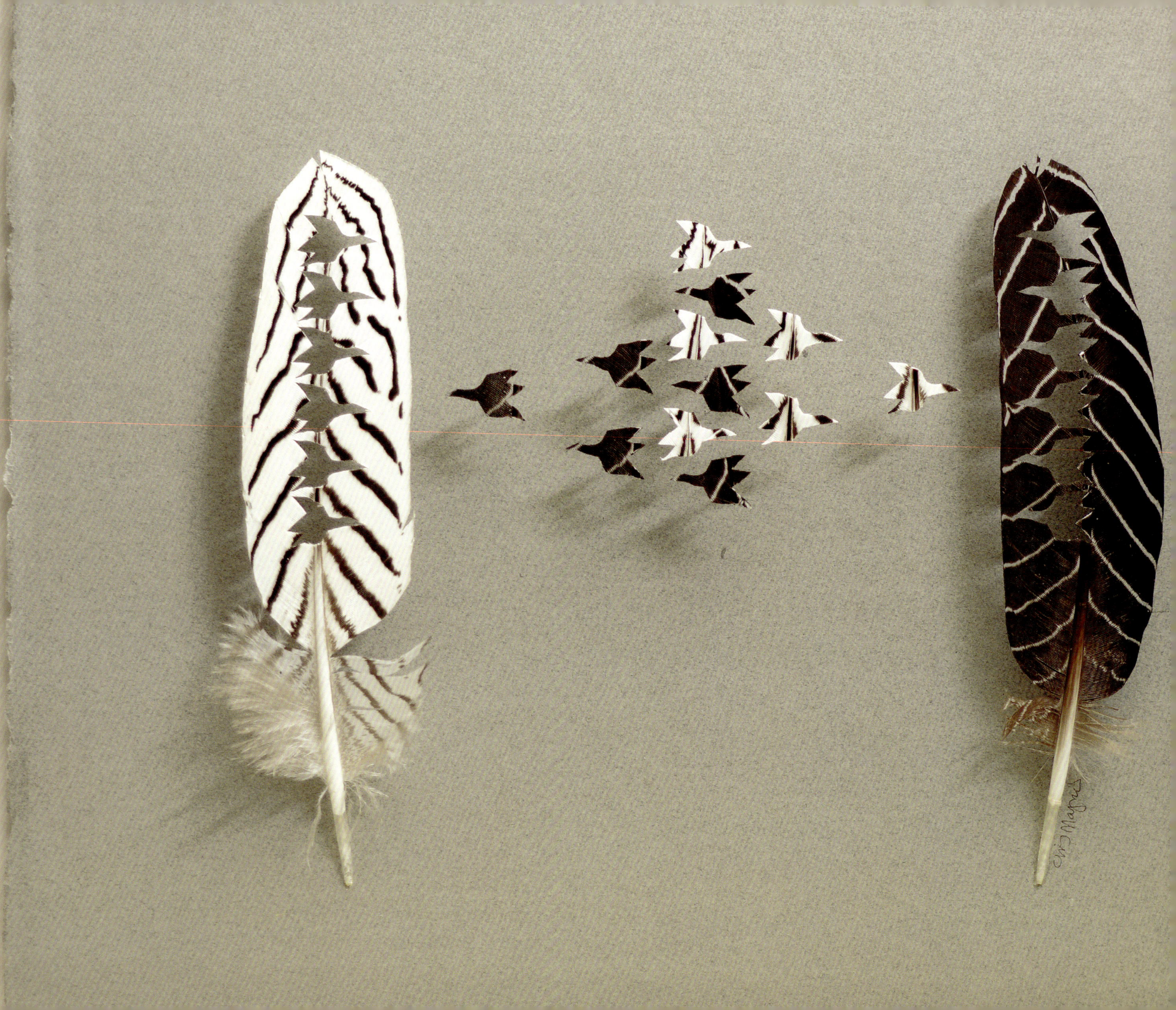

HEADED HOME 2
Silver pheasant feather and
Lewis pheasant feather
14 x 17 inches

MEANING

Why the special allure?

Each feather combines complexity and lightness into a long-lasting, intact structure that can help a bird to fly and provide warmth, coloration for display and hiding, and protection from sun, rain, bumps, and scrapes. But when we see or pick up a feather, what does it mean to us?

Birds find a lot of use for their feathers and so do we. We use them for featherbeds and down coats, decorations and fashion, fly tying, and old-style quill pens. Billions of pounds of feathers as by-products from the poultry industry find their way into animal feed, agricultural fertilizer, and landfills. Scientists and industry find more uses for feathers every year from grinding and recombining them into plastics and absorbents, to manipulation into fabric, paper, insulation, and upholstery.

In addition to these uses, feathers have always held meaning for us in our story, art, symbol, and myth.

An individual feather tells two kinds of stories. It gives clues about the bird that grew and shed it. It also allures and engages us as a symbol of beauty and flight based on our experiences, desires, and beliefs.

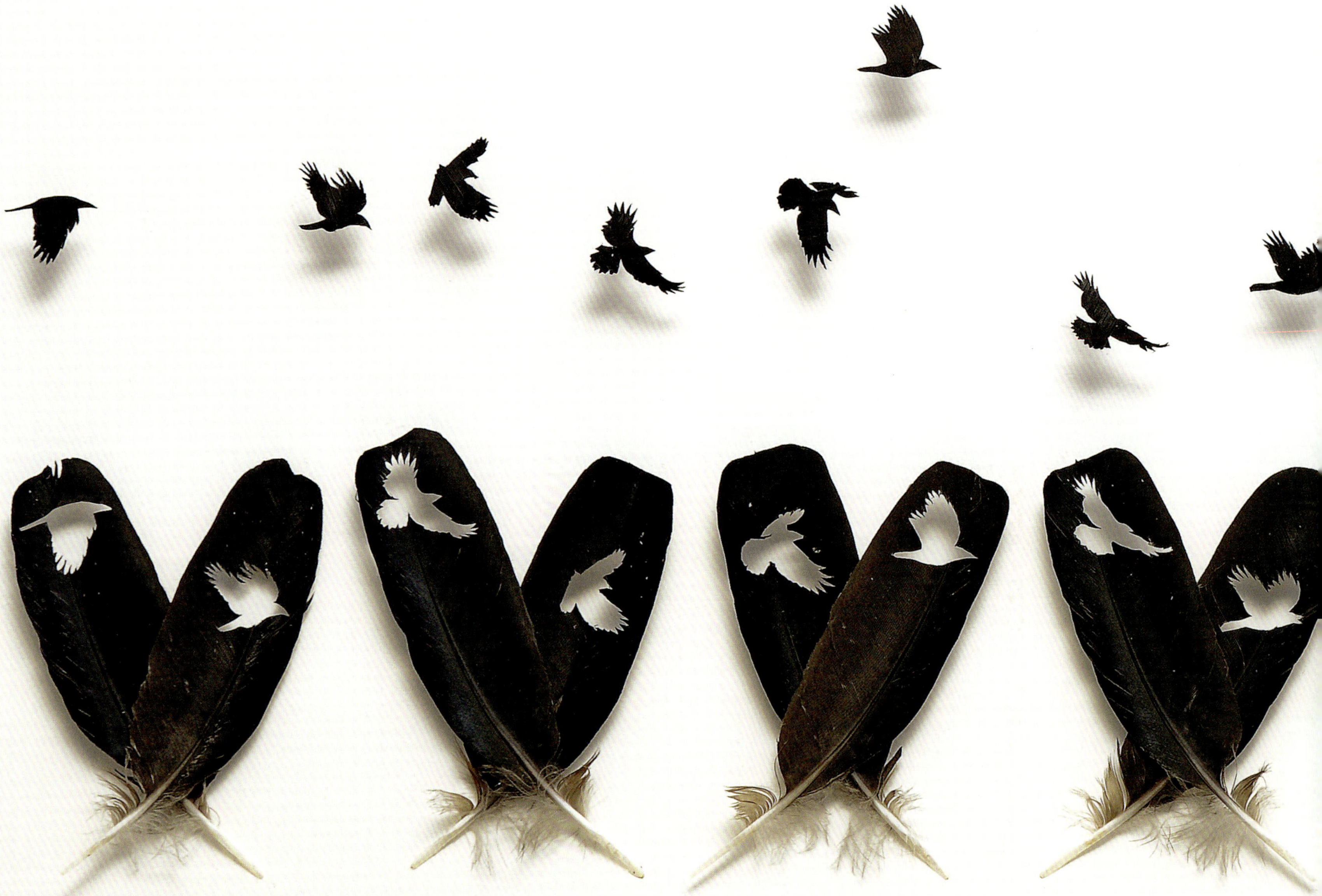

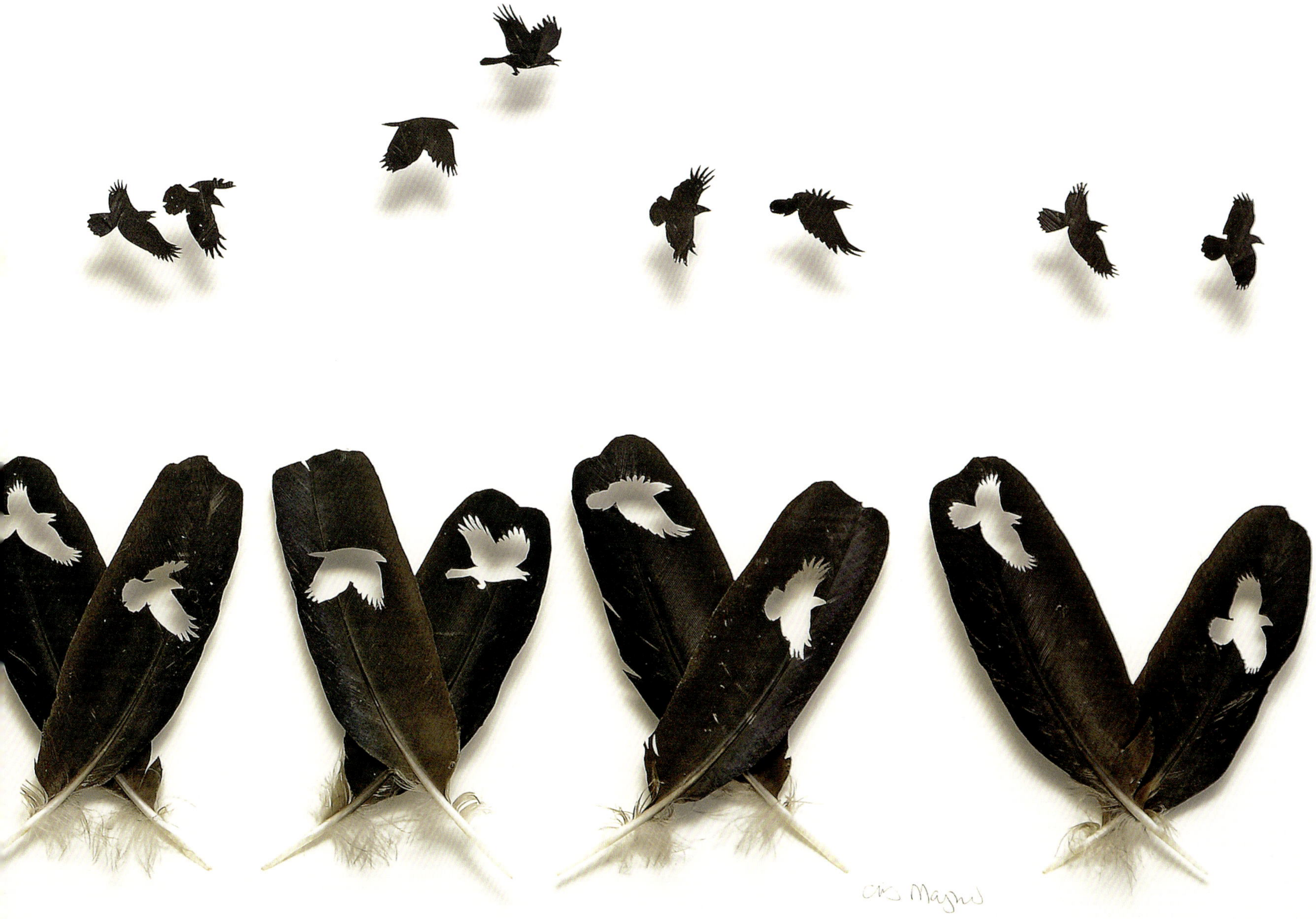

(PREVIOUS PAGE)

EVENING RUSH HOUR 2
Jackdaw feathers
14 x 32 inches

Crows experience full social lives near their families. At night, they roost in large groups in trees, arriving at dusk. By dawn, they scatter in search of food. Toward evening, they begin their journeys back to their roosting places, often miles away. The return flights converge like small streams into rivers into bigger rivers as they get closer to their roosting areas. It isn't uncommon in some areas to see 10,000 or more flying in a raggedy river silhouetted against the evening light.

Clues

When I find a feather on the ground, I consider it an opportunity to solve a mystery. What happened? I like to investigate to discover the tale of how the feather came to be in this particular place. This approach brings a refreshing attentiveness to my surroundings.

Sometimes my investigations are easy and successful. The rhododendrons outside my living room windows have become refuge for several small birds. One winter morning, from inside the house, I noticed six small songbird body feathers on the leaves and asked myself, "How did they get there?" The feathers were downy, the kind the outer body feathers cover. They were still dry so they must have appeared after the rainstorm that passed over several hours before. My guess was that the bird flew into the window. Going outside, I found no dead or injured bird lying on ground. But looking higher up, just above my head, were tiny bits of a downy feathery imprint on the window where I figured the bird had hit it. Since the birds around my house are quite familiar with their surroundings, I suspect this one got frightened, likely by my neighbor's cat, which delights in ravaging small birds. In this case, I felt I had, at least partially, solved the case of the six feathers.

At other times, I find myself with more questions than answers and I wonder about them for years. I am fortunate to have inherited a small cabin near a remote beach on the Washington State coast. When I was young, this was one of the only places in the continental United States where one could consistently see bald eagles. These days, the eagles are quite abundant. At the right time of year, several dozen at a time stand on the beach or perch in nearby trees or driftwood. They come to catch the small fish that come to spawn on the beach. Recently, during the molting season while I was walking the mile of beach from a little creek to my cabin, I counted eighty-eight eagle feathers where the high tide deposited them on the shore. The strange thing was that each one was a thin flight feather from the tip of the wing. That was it: none of the other wider feathers from farther down the wing toward the body lay on the beach. No secondary feathers, tail feathers, or wider primary feathers. That meant that well over twenty-two bald eagles had shed those feathers—two feathers from each side for a total of four feathers per bird. Only three possibilities came to mind. Either these birds were unusually synchronized in their molting, or someone picked up all the other molted feathers and didn't want the thinner ones, or some mysterious and slightly spooky force was at work.

Humans have become wired to investigate and speculate like this from millennia of hunting and gathering, attuned to our surroundings—we are just out of practice. Until recently, paying close attention to our environment often meant the difference between having a meal and having no meal. Because I get most of my meals at the supermarket in trade for money, practicing this kind of detection feels satisfying and awakens something old and deep within me.

I encourage you to do a bit of this type of detective work yourself. With a feather in hand, can you identify it? There are many questions you can ask: Is it a large and relatively straight wing feather or is it curved? Can you tell what type of bird it is from the size, pattern, or coloration? Is it the time of year you expect birds to be molting? If not, that feather is there

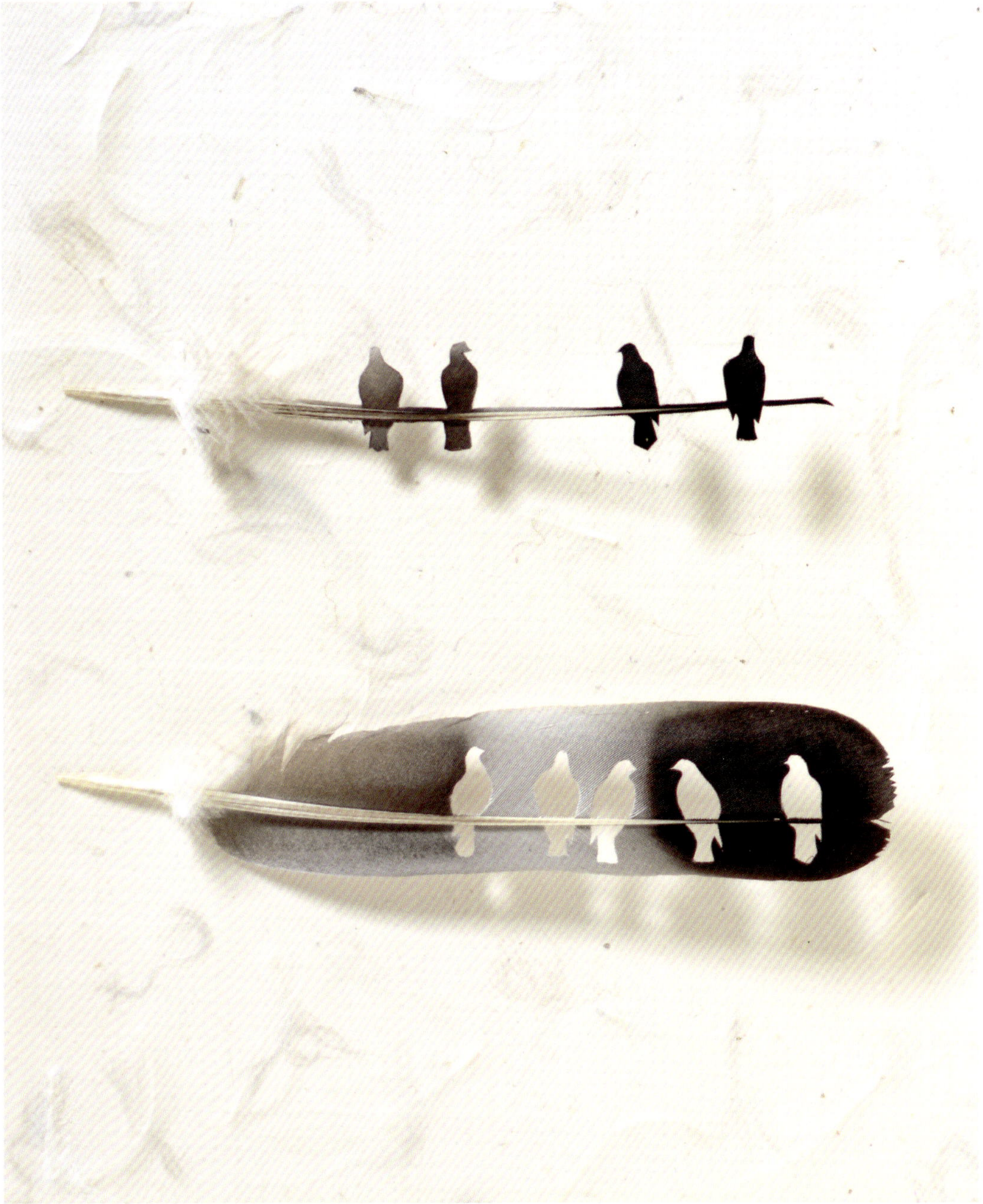

PIGEON ROOST
Pigeon feathers
11x 14 inches

The city pigeon must be the best known bird in the world. Its native home is in Europe, but it has spread to cities all over the earth.

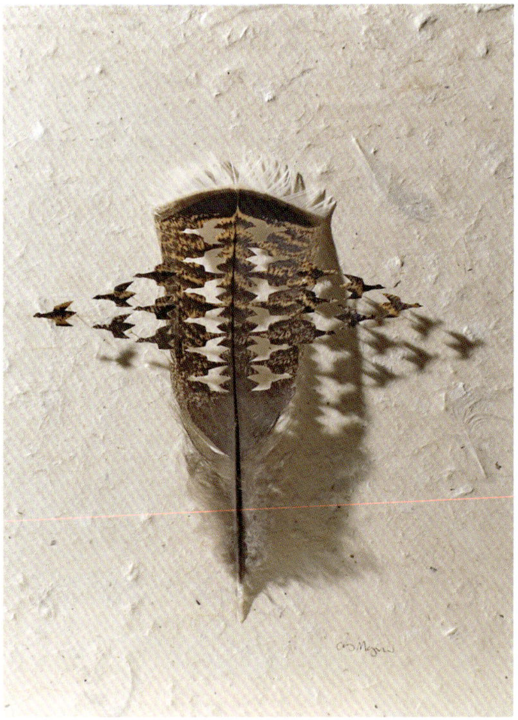

RESURGENCE
American Turkey feather
14 x 11 inches

"I am a feather for each wind that blows."

William Shakespeare

because it became separated from the bird for another reason. Are there other clues? Is the feather alone or accompanied by a bunch of other feathers? Where is it—under a bird feeder with a bunch of other feathers where a cat might have caught it? By the side of the road? On the edge of the water during hunting season? Does the shaft have bits of skin still on it? In their book, Bird Feathers: A Guide to North American Species, David Scott and Casey McFarland write, "Like all animal sign, feathers are indicators of a bird's passage through a landscape, a bit of information that we can use to piece together a larger story."

During the summer molt on our Pacific Ocean beach, I often walk among windrows of feathers at the tide line. I see gull, pelican, cormorant, and eagle feathers in abundance. I also find feathers from birds farther to sea like petrels and shearwaters.

Outside the molting season, during the winter, if I come across a feather on the beach, I take a closer look. Because of the season, the most likely reason why the feather is not on the bird is that its owner met its end in some way. Perhaps it was killed and eaten by a predator. Perhaps the feather is coated with oil, which means, sadly, the bird probably died of cold. Oily feathers to a seabird are like leaks to a boat, letting cold water soak the downy inner feathers.

While a shed feather provides us with the opportunity to heighten our sense of our surroundings, the same feather may suggest a meaning, informed by its structural perfection, lightness, and function.

Allure

Feathers are alluring. We may often bend down, pick up a feather, inspect it, and toss it to watch it float away or spiral back down to the ground. We might even keep

EAT 2
Female argus pheasant feather
17 x 12 inches

If a bird eats insects and when the bird dies, insects eat the bird, a cycle becomes complete.

it for a while as a memento. For me as an artist and biologist, the physical wonder of the feather is enough to pull me into its charm. But there is more to it. Feathers and images of feathers have found their way into symbols in cultures dating back millennia to represent ideas humans hold as important. I began wondering whether something more than physical allure makes me fascinated with feathers, something under my skin. I decided to seek help from a local tattoo shop.

Marco Hernandez owns Lit Fuse Tattoos and has worked as a tattoo artist for many years. In addition to having been asked to tattoo feathers, he has done birds and other animals, Celtic motifs, Japanese styles, tribal designs, and the proverbial anchor. This is a place where myth and

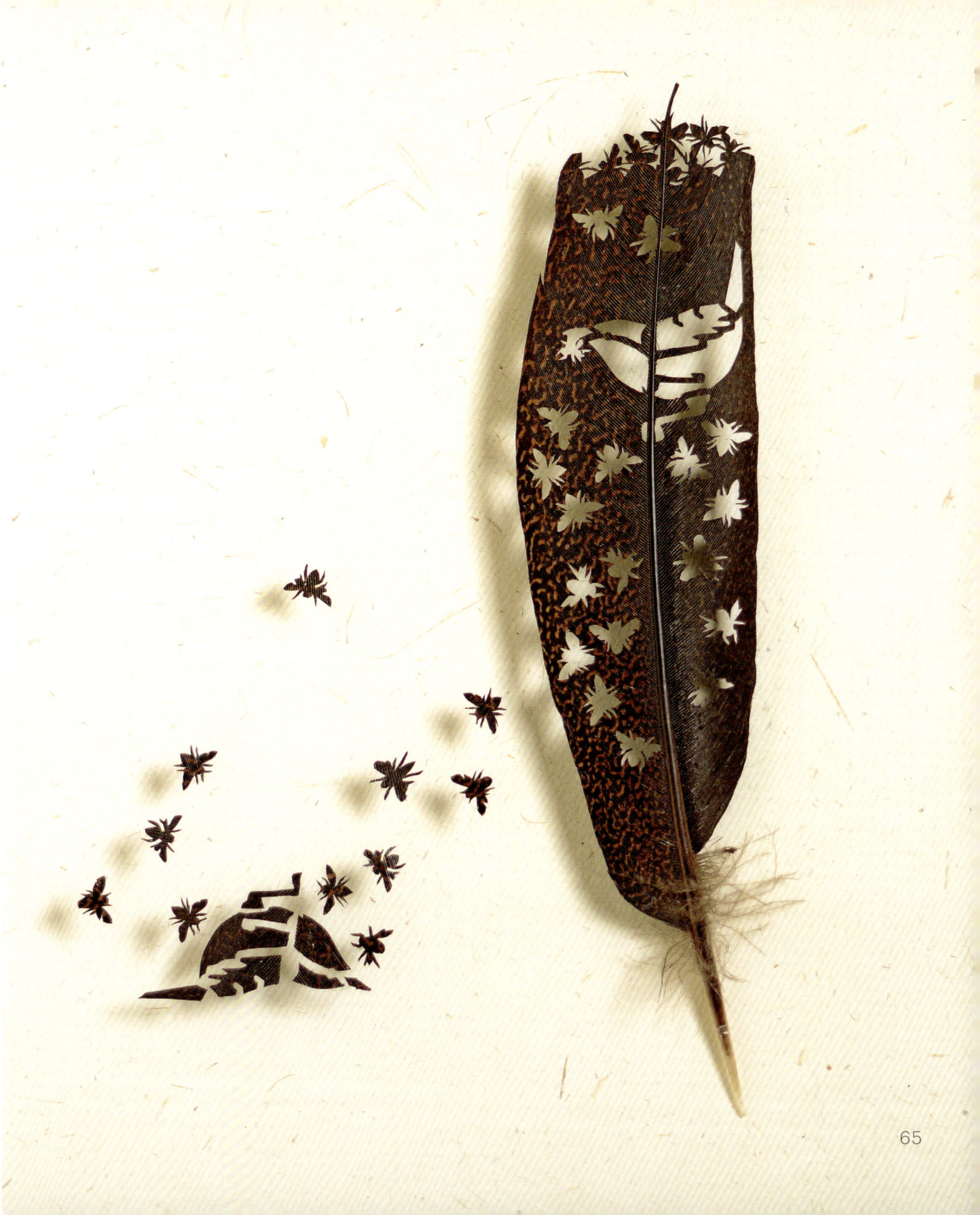

meaning are best discovered in person these days. It seems that every corner and strata of society is tattooed in some way, and feathers are one of the most popular themes. I felt I had come to the right kind of place. I mean, these designs are inked into the skin—permanently! I like to think a person chooses a feather design with the attitude of Angelina Jolie, "A tattoo is something permanent when you've made a self-discovery, or something you've come to a conclusion about."

I went into the back room and watched Marco using his tattoo machine to make a colorful design on a man's arm. Marco invited me to talk with him about feathers and tattoos, but I felt uncomfortable distracting him from concentrating on his work. I said I would wait in the lobby.

On the counter, I found a book of photographs taken of the artist's original ink designs. These were the artist's pride and joy and reflected his style. Also,

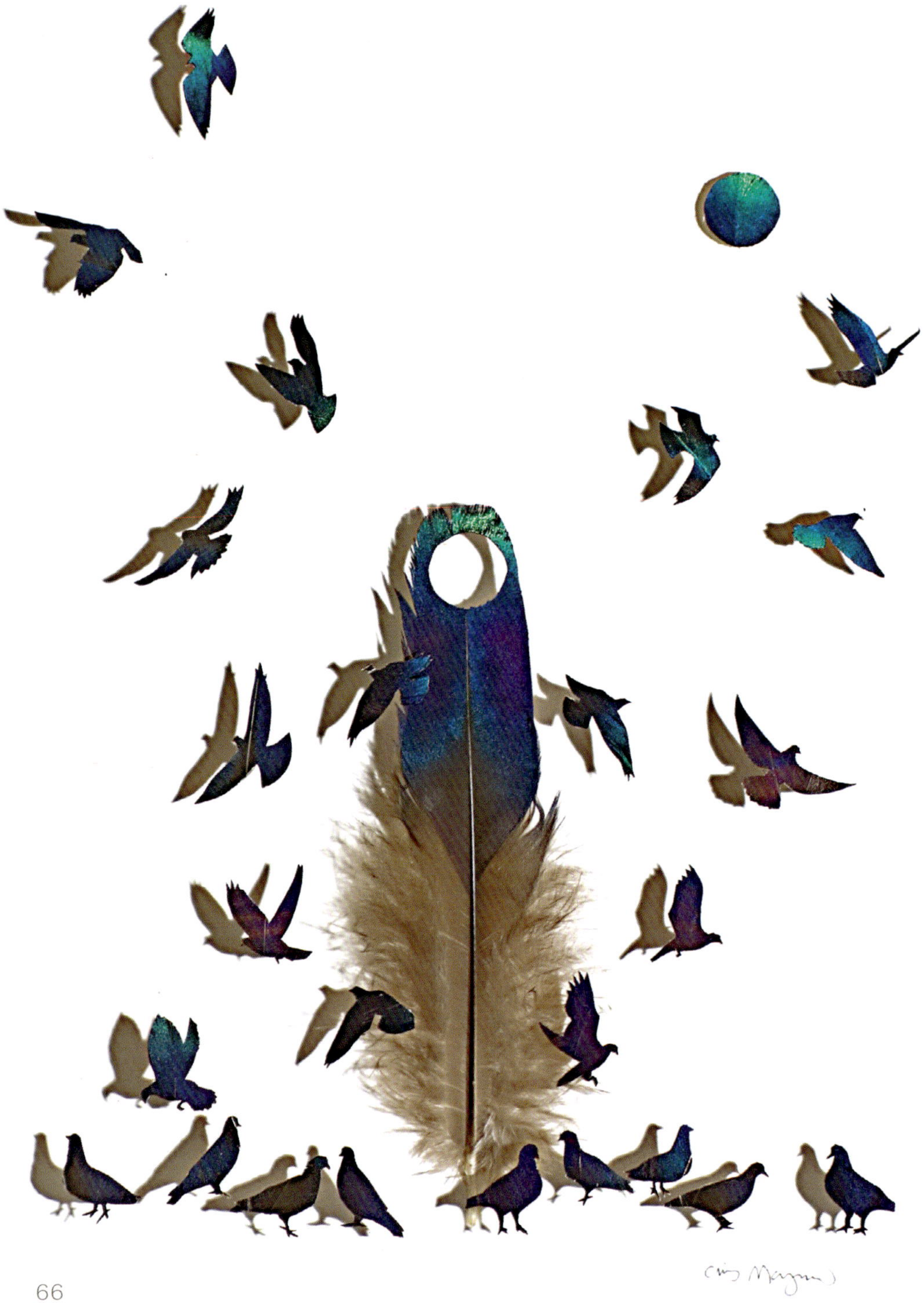

PIGEON PARK
Impeyan pheasant feathers
14 x 11 inches

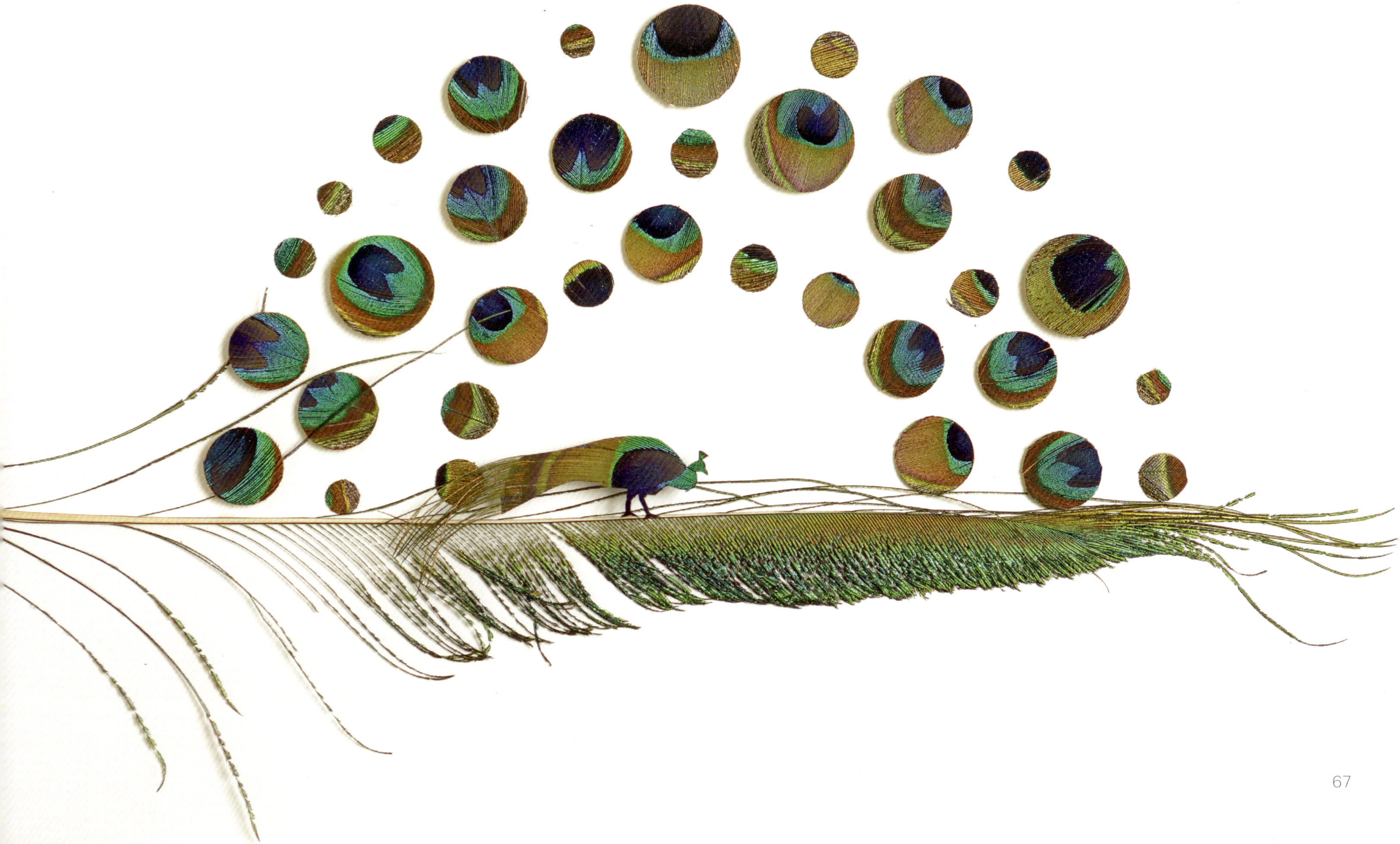

PEACOCK ATTRACTION
Peacock feathers
11 x 14 inches

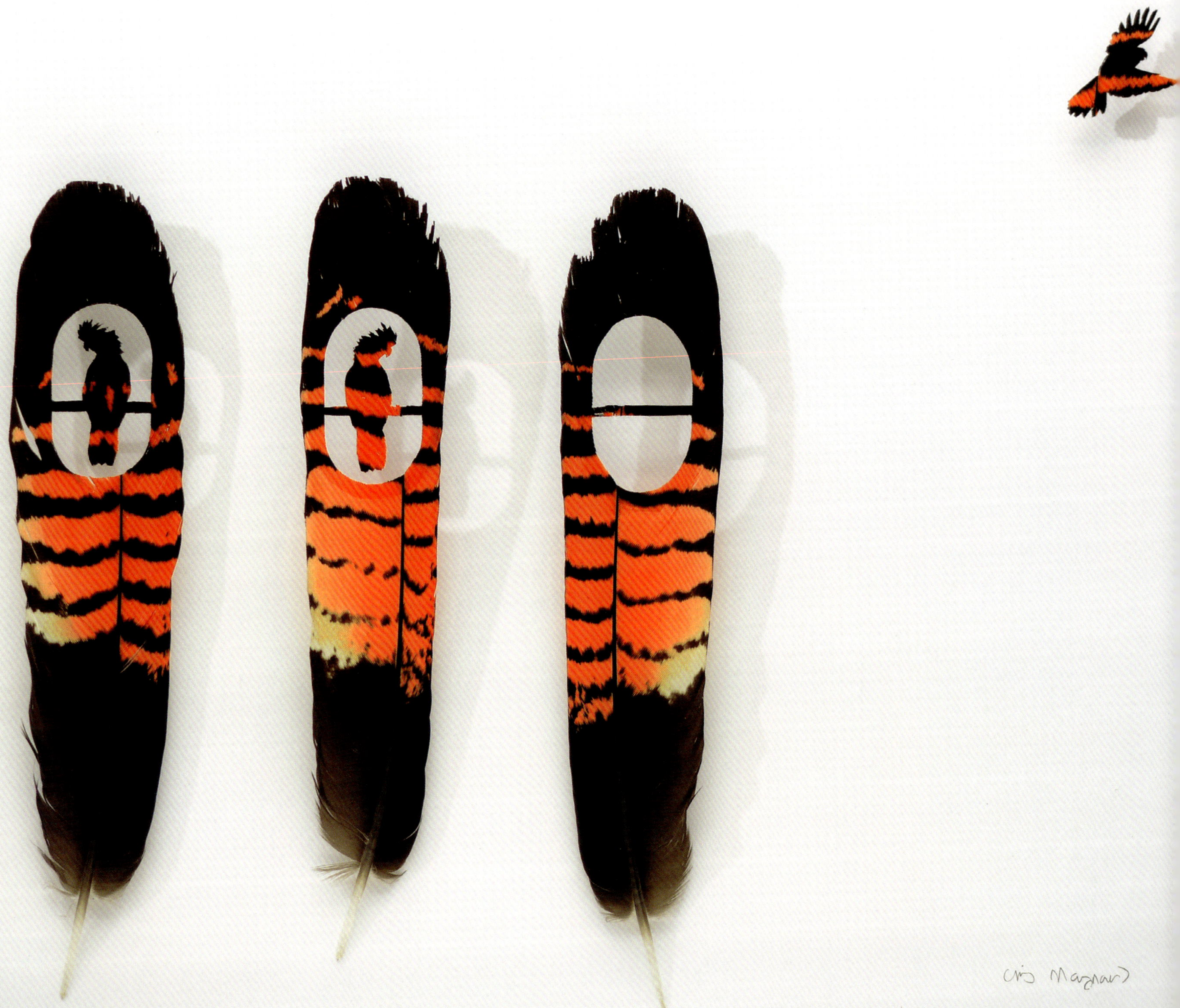

CHOICES

Red-tailed black cockatoo feathers
19 x 23 inches

Shall I go or shall I stay
The door is open, fly away?
But it's safe, my food is here
Choices are not always clear

as in hair salons that have hair magazines on their tables for waiting customers to peruse, this tattoo shop displayed standard designs in books to help customers make their choices. I leafed through the books and scanned the posters of tattoos on the walls. As I listened to the buzz of Marco's tattoo machine, I flipped through many different motifs, themes, and colors.

Many of the feather tattoos in the books were accompanied by quotes that gave the tattoo meaning. Here are a few that were often repeated: "One day I'll fly away", "Feathers appear when angels are near", "Drifting on a memory", "Hoping with a fragile hope", "You can't kill a man when he's got no hope" (sadly accompanying the tattoo of a fallen feather).

The quote, "Drifting on a memory" was accompanied by a soft, curled body feather. I thought of the beginning of the movie *Forrest Gump* when a soft-looking curved white body feather drifts on a whisper of wind, light, free from the world's cares in an effortless, seemingly random but perhaps guided flight. One senses from this image that everything will turn out well and that getting there will even be easy. It sets the tenor and theme for the entire movie. Of course, the music helps.

In the shop's design books and posters, the major theme in most of the feather tattoos was derived from the fact that feathers help a bird fly. Feathers are, in a sense, symbols of flight, of something we long to do. To us, flight often means transcending difficult circumstances, or sometimes, it means simply taking delight in life. In their random trajectories, detached feathers suggest to the imagination our human course of life, neither fully grounded nor fully ascended, but simply emblematic of hope itself, suspended as bridges between here and heaven, earth and sky.

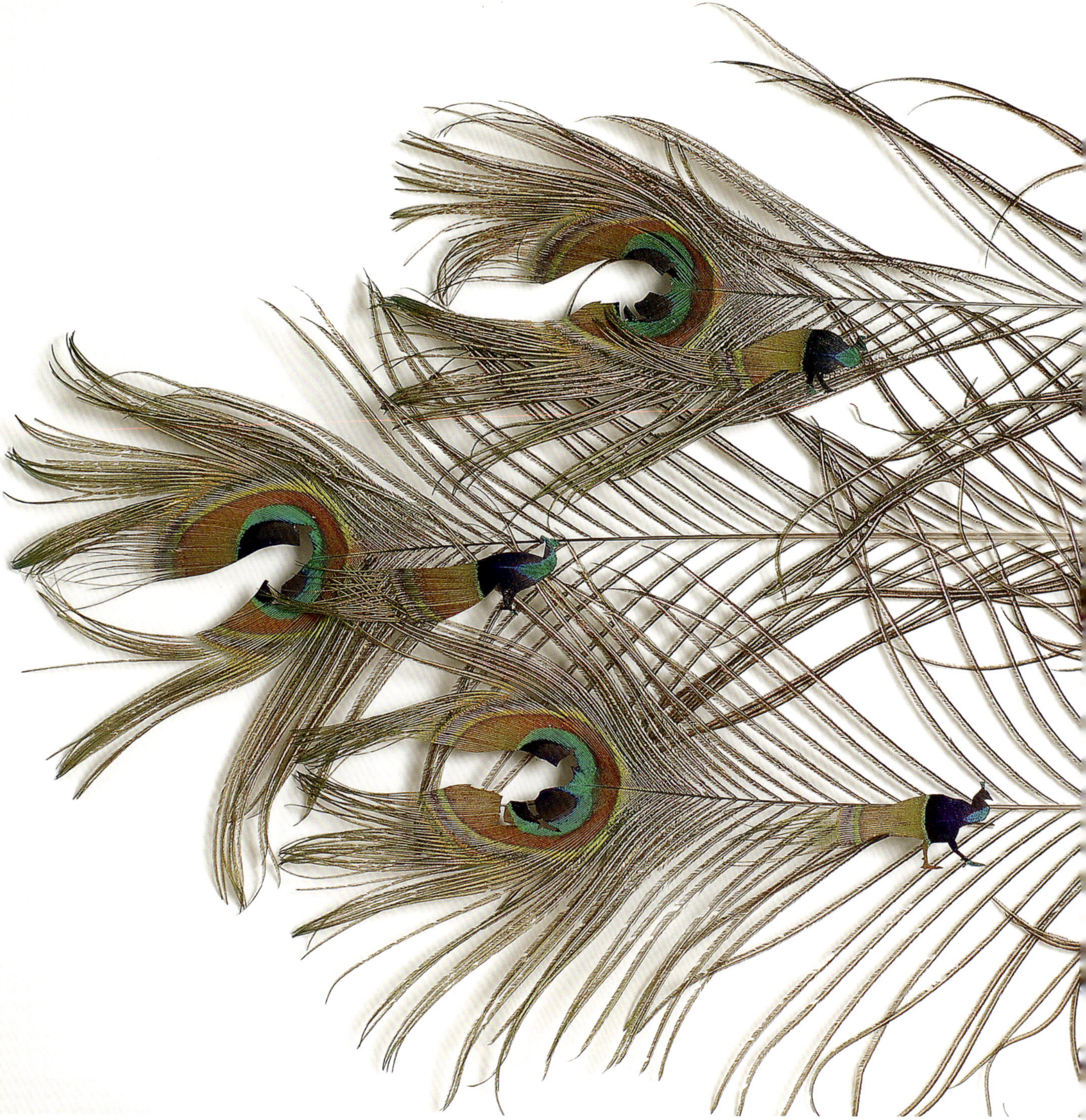

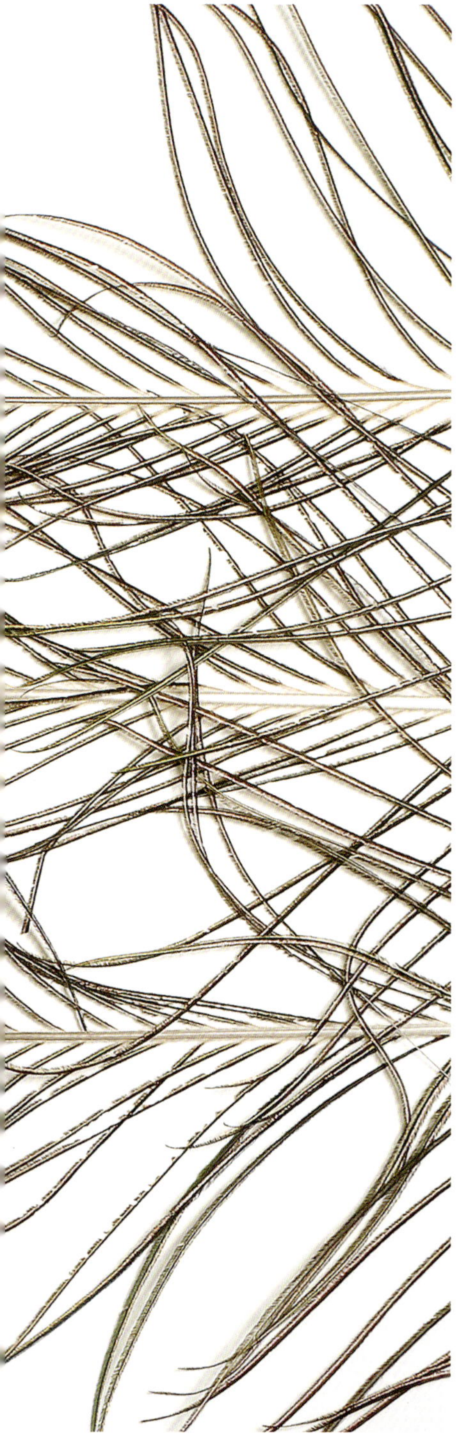

BEAUTY ON THE MOVE
Peacock feathers
14 x 21 inches

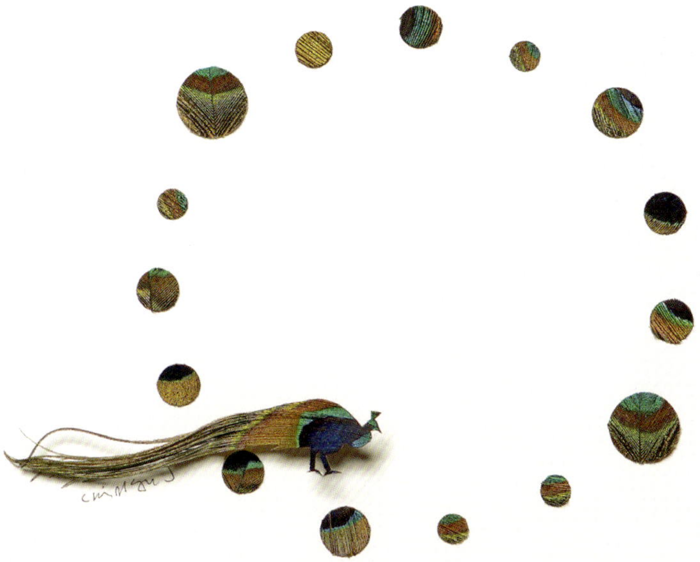

PEACOCK SQUARE
Peacock feathers
12 x 12 inches

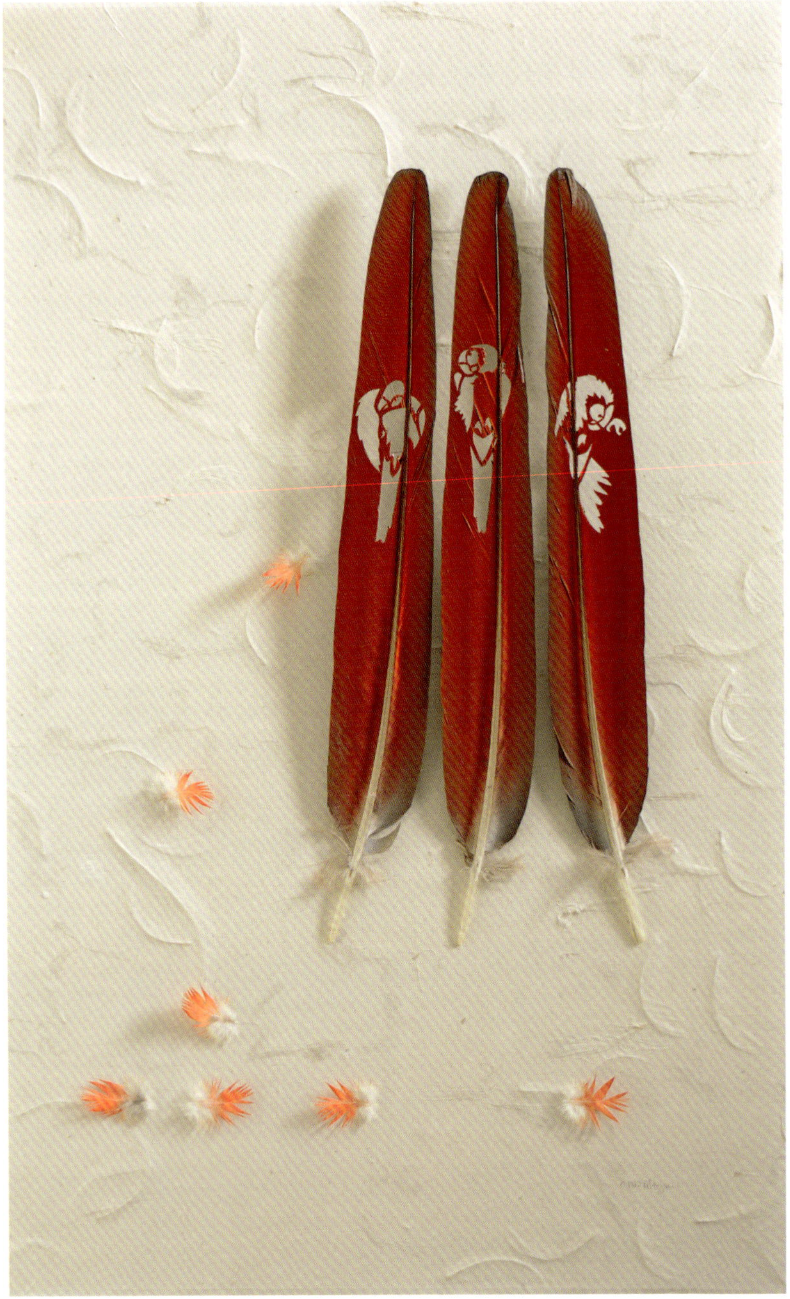

One quote I read found me perturbed. It had to do with the word "fragile." It said, "Everything beautiful is fragile." This concept seems mistaken to me. Even fashion designers who employ feathers for their perceived "feminine" and delicate qualities know that their feathers are tough. But because feathers are also light and wispy, they are sometimes seen as fragile. I often have to convince people that my artwork is sturdier than it appears. Feathers, though close to weightless for flight, have to be sturdy in order to keep a bird warm and dry, to perform at freeway speeds in the air, and to protect their owner from scrapes for up to a year until they are replaced with new feathers.

MACAW PERCHERS
Green-winged macaw tail feathers and scarlet macaw body feathers
22 x 14 inches

SWIMMING DUCKS, DETAIL
Mallard speculum feather
11 x 14 inches

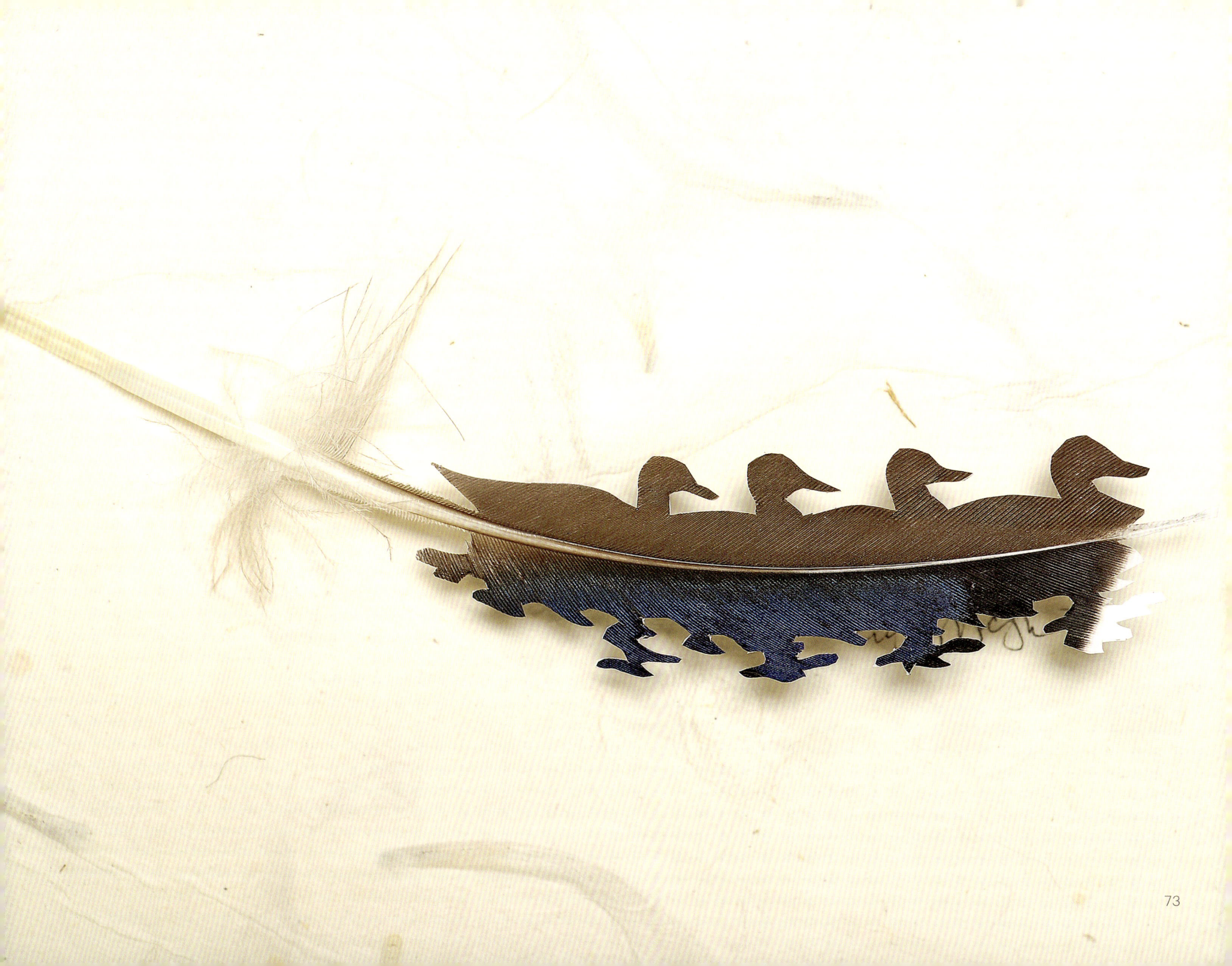

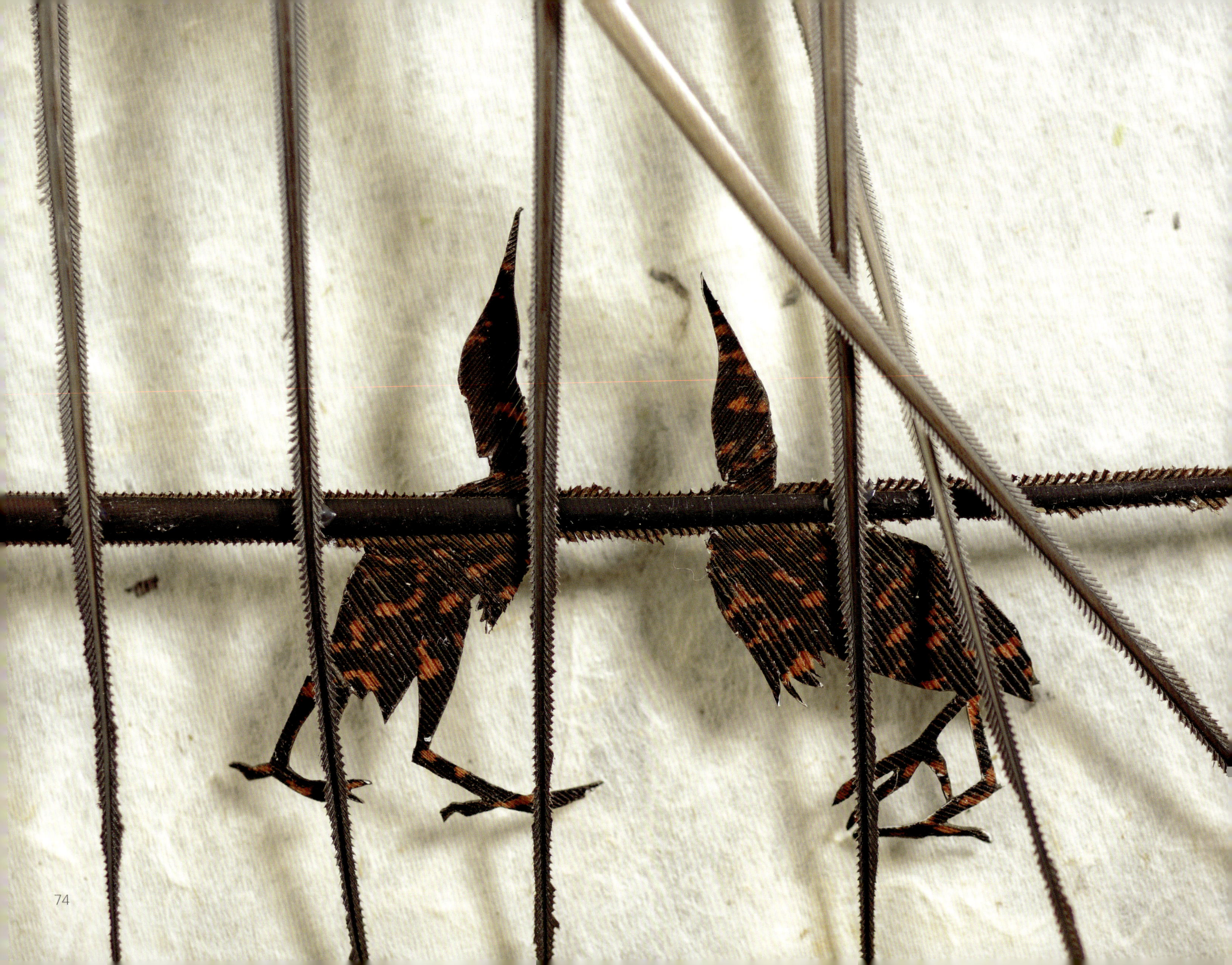

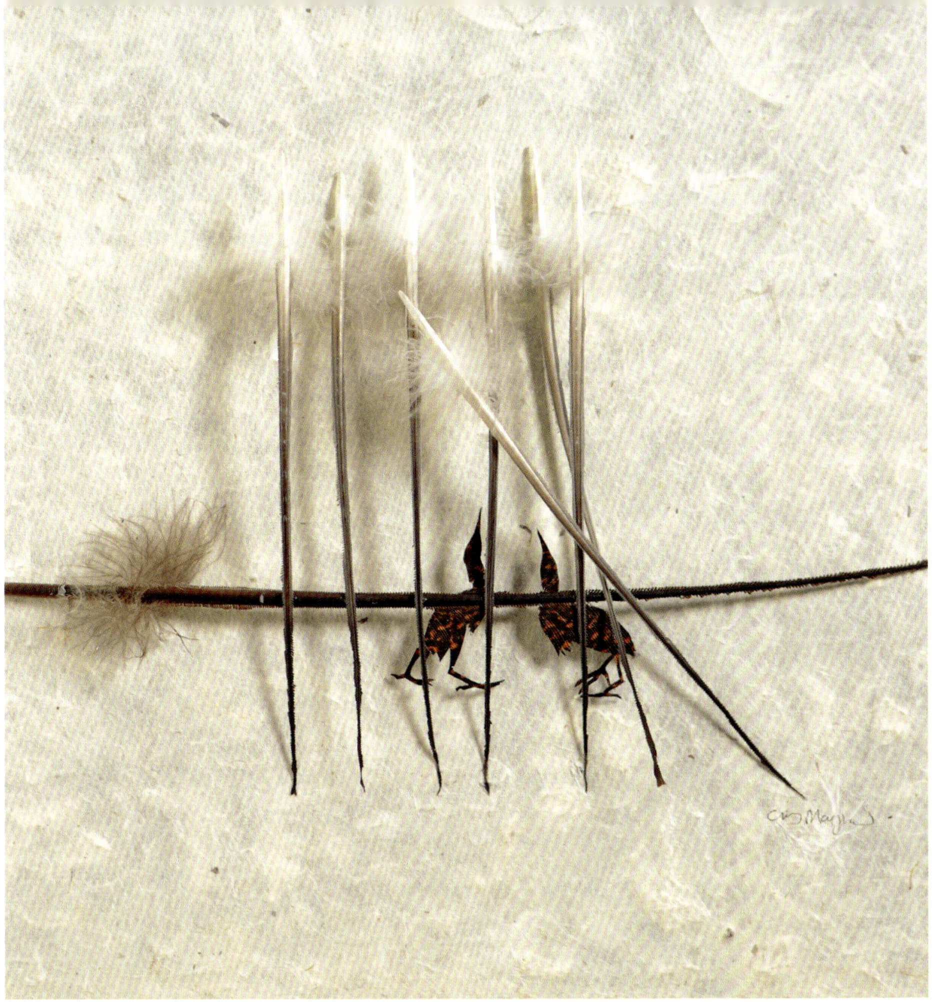

BITTERNS 2
Argus pheasant feather and
pigeon feathers
11 x 14 inches

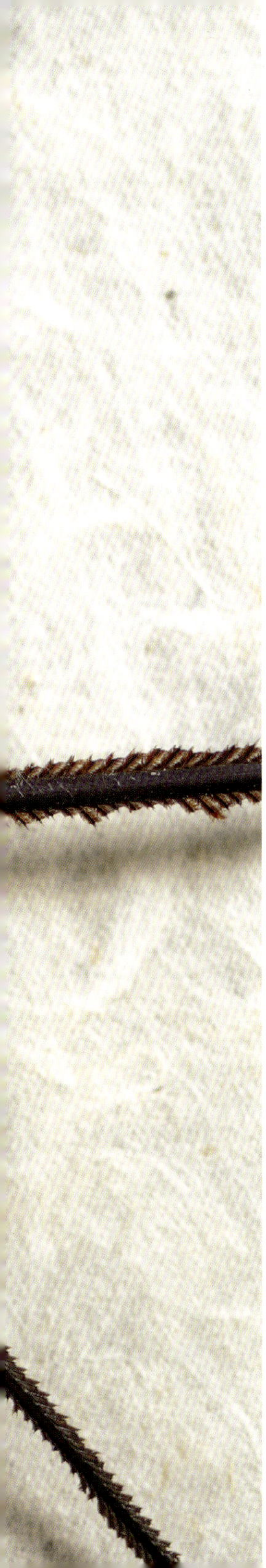

BITTERNS 2
(detail)

Of course, a lot of artistic license was evident in the tattoo designs. Not very many feathers had a firm base in realism or biology. These images were mostly idealized feathers, each image adjusted to match the meaning of its accompanying quote.

Perhaps my own personal discovery of how feathers can be symbols is obscured by my training in biology and experience with real feathers. So when working, I have needed to suspend my knowledge of science and remind myself that any art, including mine, is a representation of an idea, not the real thing.

The other major feather theme not related to flight was found in tattoos of peacock display feathers (which are not actually tail feathers but rather emerge from the peacock's back). I found that artists who ink peacock feather designs usually use colors that closely match a real peacock feather. (Today, tattoo inks are not just dark blue and black from carbon and other natural ingredients; they come in many bright premixed colors that often include the same color-producing metals such as copper, mercury, lead, arsenic, cobalt, titanium, and cadmium used in oils and acrylics of traditional fine arts.) These artists stylize their designs, but they still make them easily recognizable as peacock feathers.

Peacocks compete only with the more obscure great argus pheasant for the most lavishly designed display feathers. Otherwise, these feathers are useless; for a peacock, these colorful feathers are only for love.

We associate a curious collection of myths and qualities with peacock display feathers. These feathers seem to have a complicated relationship with cultures around the world. For example, while many cultures associate them with fortune, compassion, kindness, patience, openness, resurrection, immortality, and renewal, some (especially around the Mediterranean) think of peacock feathers as bad luck. Some believe they are a symbol of vanity. Others say never to bring a peacock feather to a wedding. And some even go so far as to say you should never ever sleep on a bed of peacock feathers or you will die.

I began looking for references to particular birds in the tattoo shop's design books. Besides peacocks, I recognized eagles, owls, swallows, phoenixes, crows, and ravens. All of these feathers were displayed as representing flight in some way. The eagle feathers tended toward strength; owl feathers were a bit of a mystery but drawn more realistically; swallows showed freedom and exuberant flight; colorful phoenixes, of course, were rising from flames, symbolizing transformation. Ravens and crows' black feathers often connoted more complex meanings. One black feather was pictured on the ground with a quote, "Where the angel and devil meet."

The crow feather tattoos reminded me of a letter I once received. It contained a picture of the sender, a young woman who had a new tattoo on her back of nine crow feathers with crow silhouettes cut out and flying above the feathers. It was copied from my shadowbox design entitled "Evening Rush Hour," which you can find on page 60-61. Crows are familiar, intelligent, and social. We have opportunities, in the cities and urban landscapes where most of us live, to observe closely and interact occasionally with these smart birds. So I expect some of the crow tattoos I was looking at in the book represented unique meanings based on personal experience.

When Marco finished with his client, he asked if I would like a tattoo of a feather. I said, "No" but thought that if I did, I would get two small ones—flight feathers out of sight somewhere, one on each side, so not to create unbalance in me like a clipped goose.

I think Marco is a bit like me, focused on creating his art and leaving the identification of meaning to his clients. He pointed out that

CROW CAW
European crow feather and silver pheasant feathers
11 x 14 inches

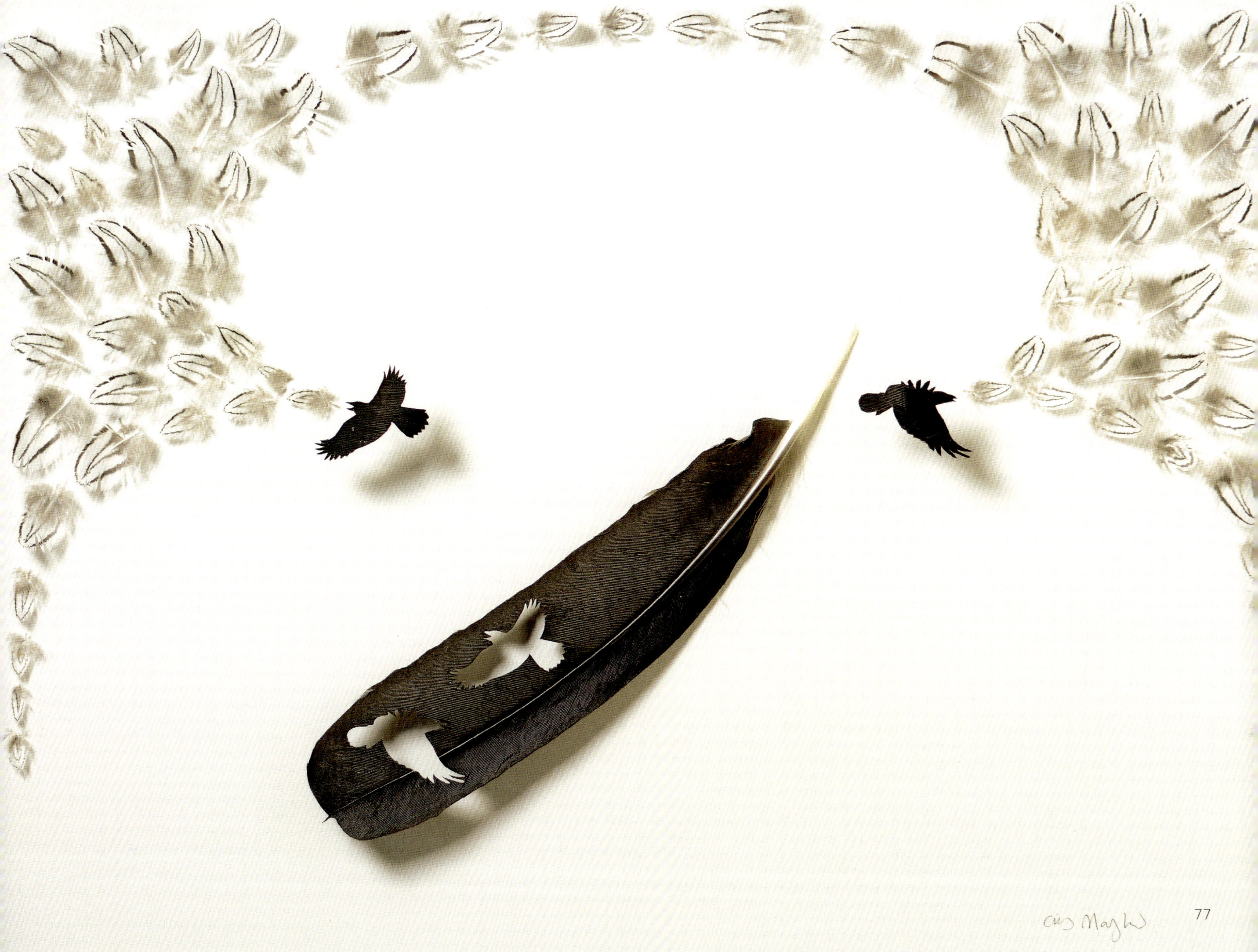

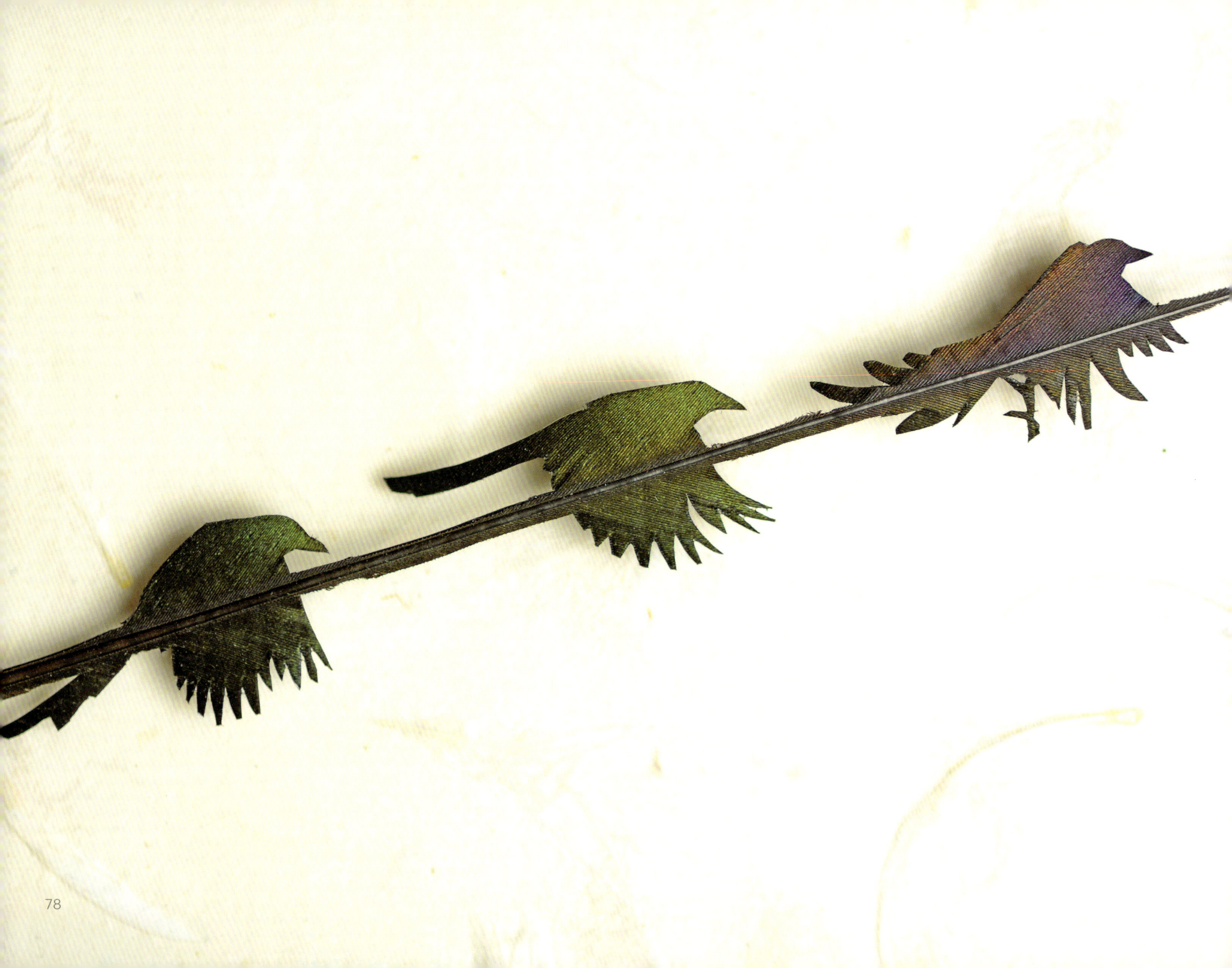

MAGPIES
(detail)
Eurasian magpie tail feather
16 x 20 inches

Feathers are seen as a glimpse into possibilities only real to us in our dreams.

while his clients express attraction to loose themes of flight and lightness, the main theme is "There is no main theme." He says, "Feathers mean something individually to each person. To many, they just like how the feather images look." I suspect, though, that an unconscious sense of meaning lurks under both of our artistic creations and is more influential and rich in our art by being unrecognized, unnamed, and unseen.

At home that evening, I was curious to see whether the placement of these tattoos had anything to do with the feather's meaning or purpose. On a popular photo-sharing website, I searched for "feather tattoo" and categorized images of 400 feather tattoos according to where they were on the body. The only definitive conclusion I could make was that more peacock designs were placed where the owner couldn't easily see them, yet they could be seen by others. This made sense because display is what peacocks do.

I expected that upper backs and shoulders (corresponding to a wing's location on birds) would sport more feather tattoos than other places on the body. But I also found just as many that were inked on wrists and sides, legs and feet. In my experience, women are more drawn to feathers than men for whatever reason: 95 percent of the 400 images I found were on women.

The most popular feather tattoo image was one of birds flying out of a feather, similar to the image on page 28. The most popular saying accompanying feather tattoos was, "One day I'll fly away." It is no wonder that with this kind of yearning, feathers are seen as a glimpse into possibilities only real to us in our dreams.

We make sense of our world in many ways: art, science, religion, philosophy, our own experiences, myth. Even though feathers are perfect structures by themselves, art and using feathers in art serves to add meaning for us. Artists use their mediums (in my case, feathers) to direct their viewers to ideas to which they can relate. The feathers themselves are just innocent participants of our mind's workings.

I hope you put this book down with an understanding of why feathers are important to me. I'm drawn to them because of their beauty, their strength, and the way they seem to draw in our imaginations. I've seen the way people often approach my artwork: as if they feel an unspoken connection, a familiar comfort, with the feathers I use. This may just be personal, but I like to believe it has to do with something deeper, something about the interconnectedness of the whole animal kingdom. That is my hope for this book: that you put it down feeling more connected to the world birds live in, because, after all, it's your world too.

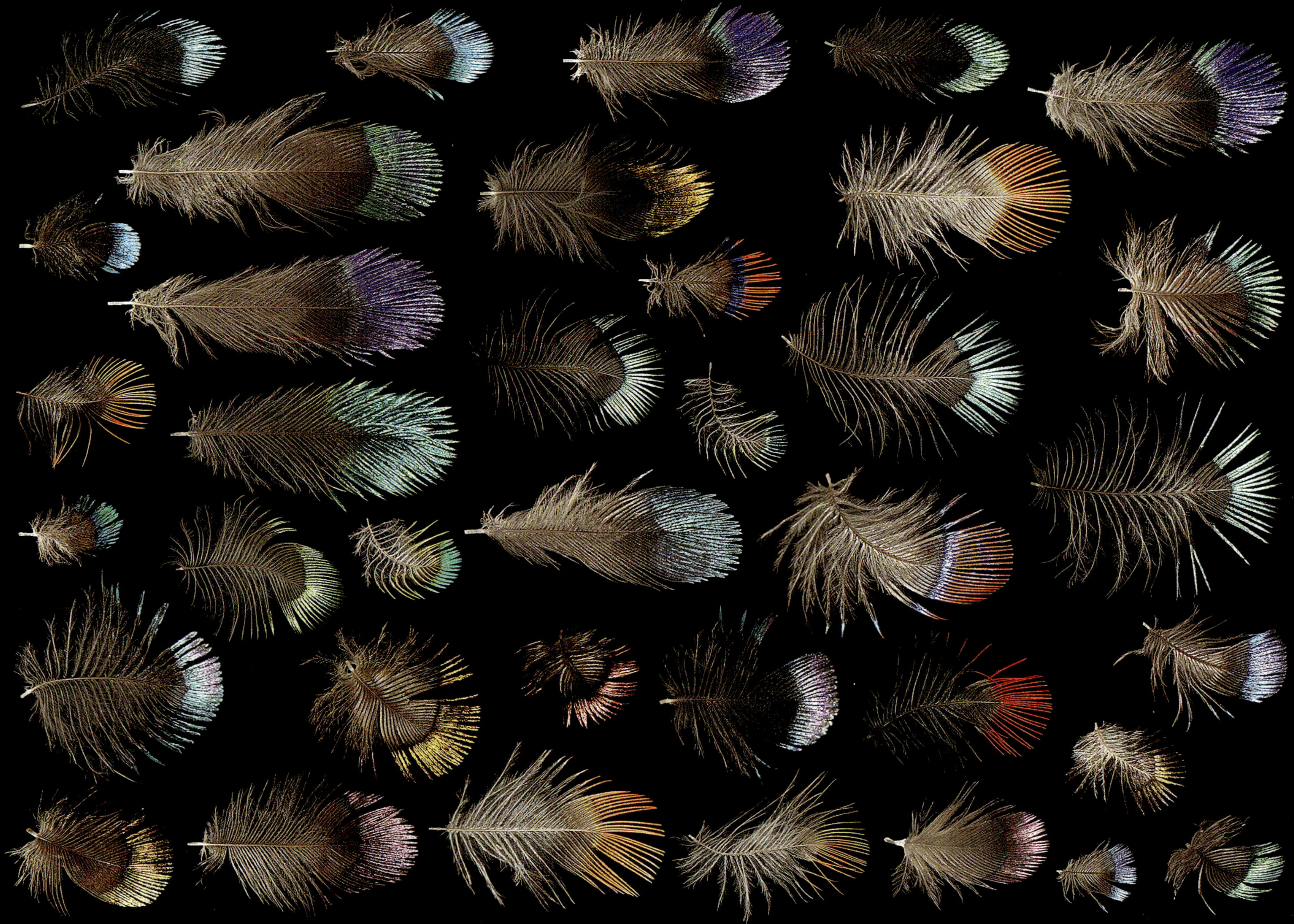

A FEATHER

Photography Appendix

Before I began cutting and carving feathers, I photographed them. At one time I was seriously considering going a more scientific route, beginning with photographically cataloguing the large bird wing collection at the Burke Museum in Seattle. My creative focus on beauty prevailed though, for which I am grateful. The perfection of each feather beckons and beguiles; an accomplishment of form to be enjoyed with minimal human artistic interference. So I still find delight in capturing and sharing the beauty of feathers through photography.

The metallic color reflections in feathers dazzle our eyes, but only if the light is right.

I carefully direct light onto a feather so the brightest of the color reflects into my eyes when looking through a camera viewfinder. I take a picture of a single feather at a time, then combine many pictures to make what can be blown up into a huge photomontage of, say, turkey feathers or sunbird feathers.

AN ASSORTMENT OF AFRICAN SUNBIRD FEATHERS
photograph

Photographing these some as small-as-a-grain-of-rice feathers from eight species of sunbirds required careful breathing to avoid sending the feathers airborne with the slightest puff.

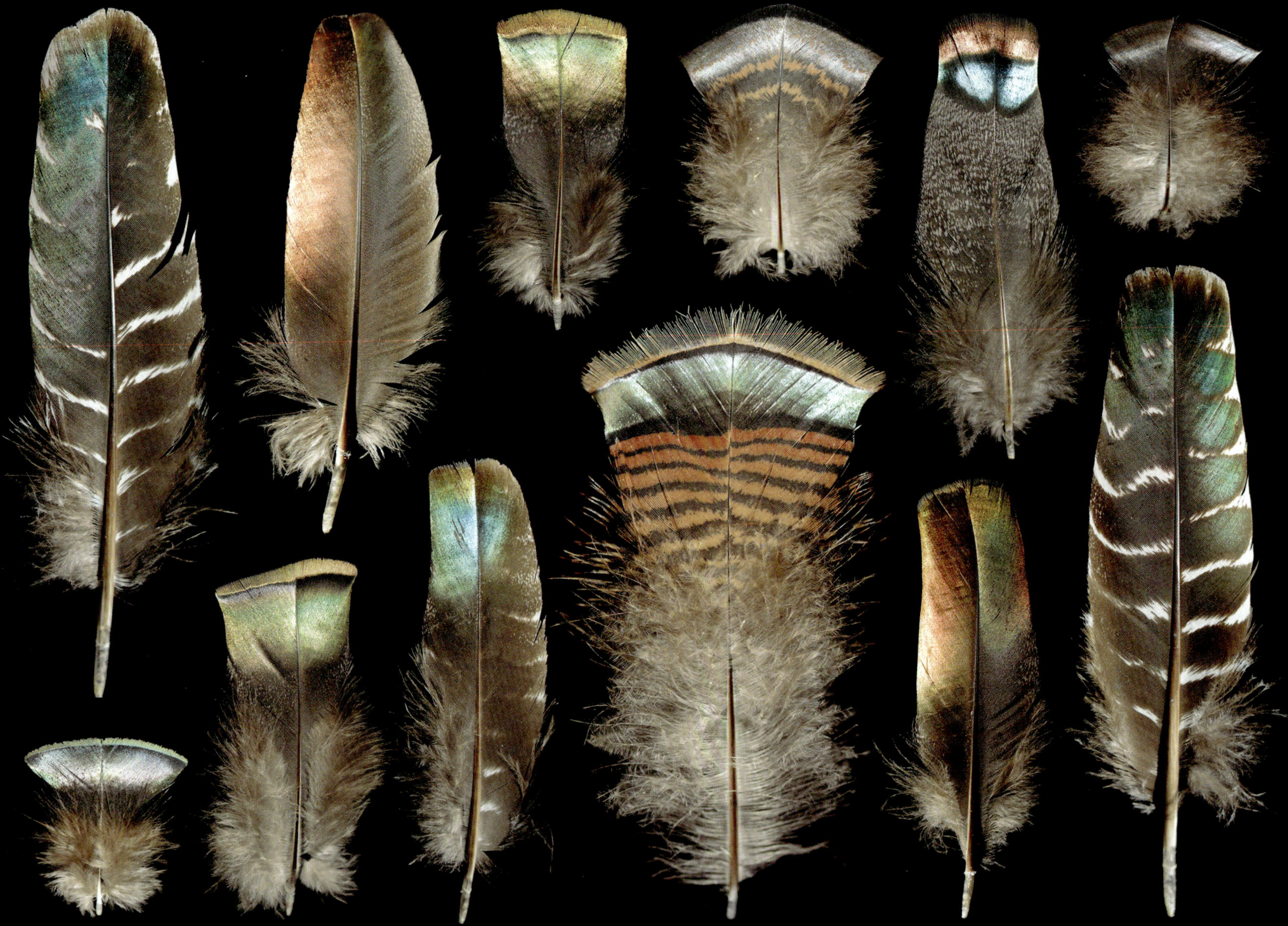

(LEFT)

A TURKEY ASSORTMENT

photograph

Though many variations of the American turkey roam the wild and our farms, only two species of turkeys exist: the American and the ocellated, which lives in Central America. These feathers are mostly from a captive ocellated turkey which may have interbred with an American turkey.

A PEACOCK ASSORTMENT

photograph

When people raise a creature for generations, a lot of interbreeding goes on. The India blue peacock has been bred into ten or so varieties, but actually only two species of peacock exist. The photo shows varieties of India blue feathers and some of the spectacular tropical green peacock feathers.

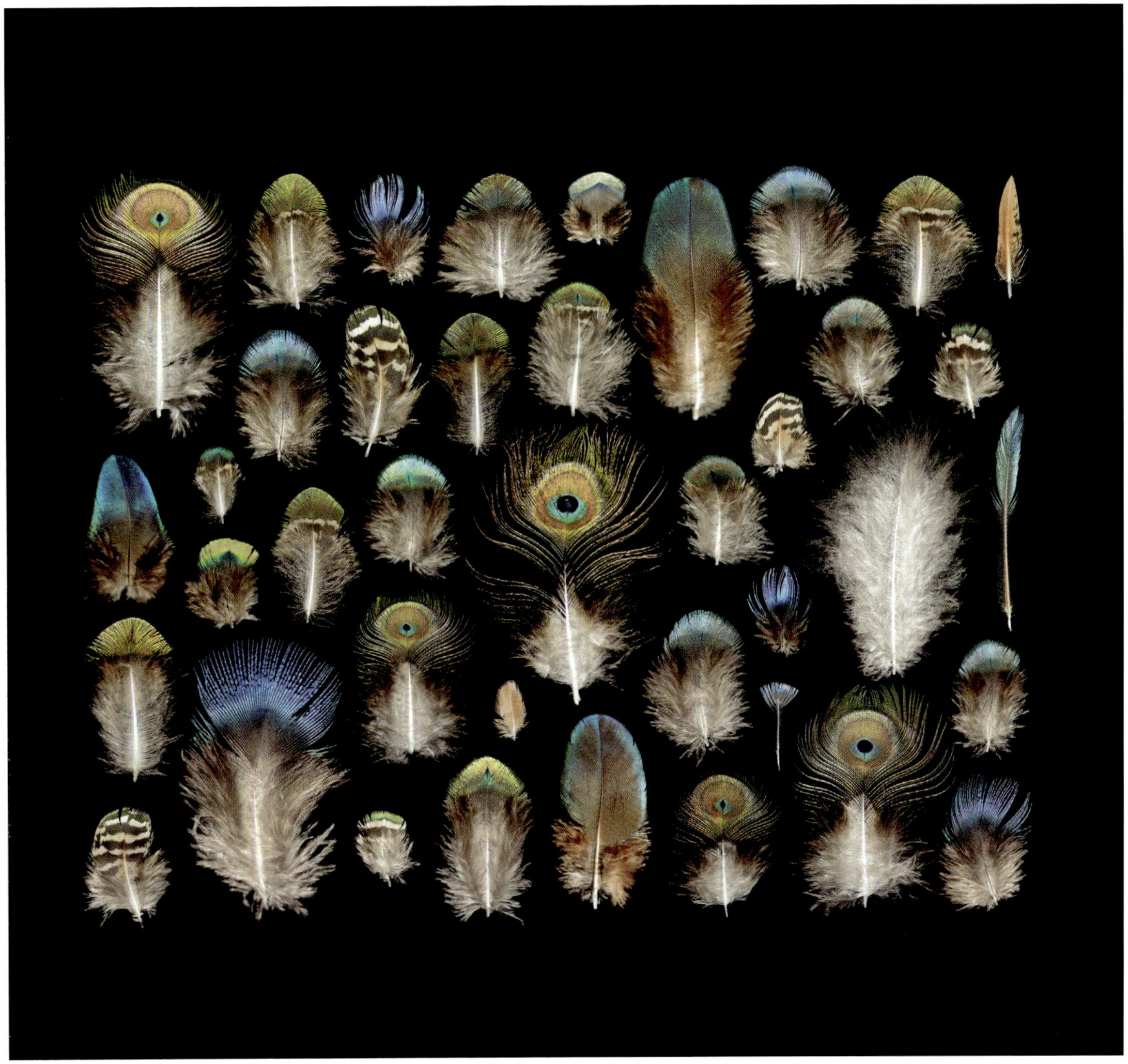

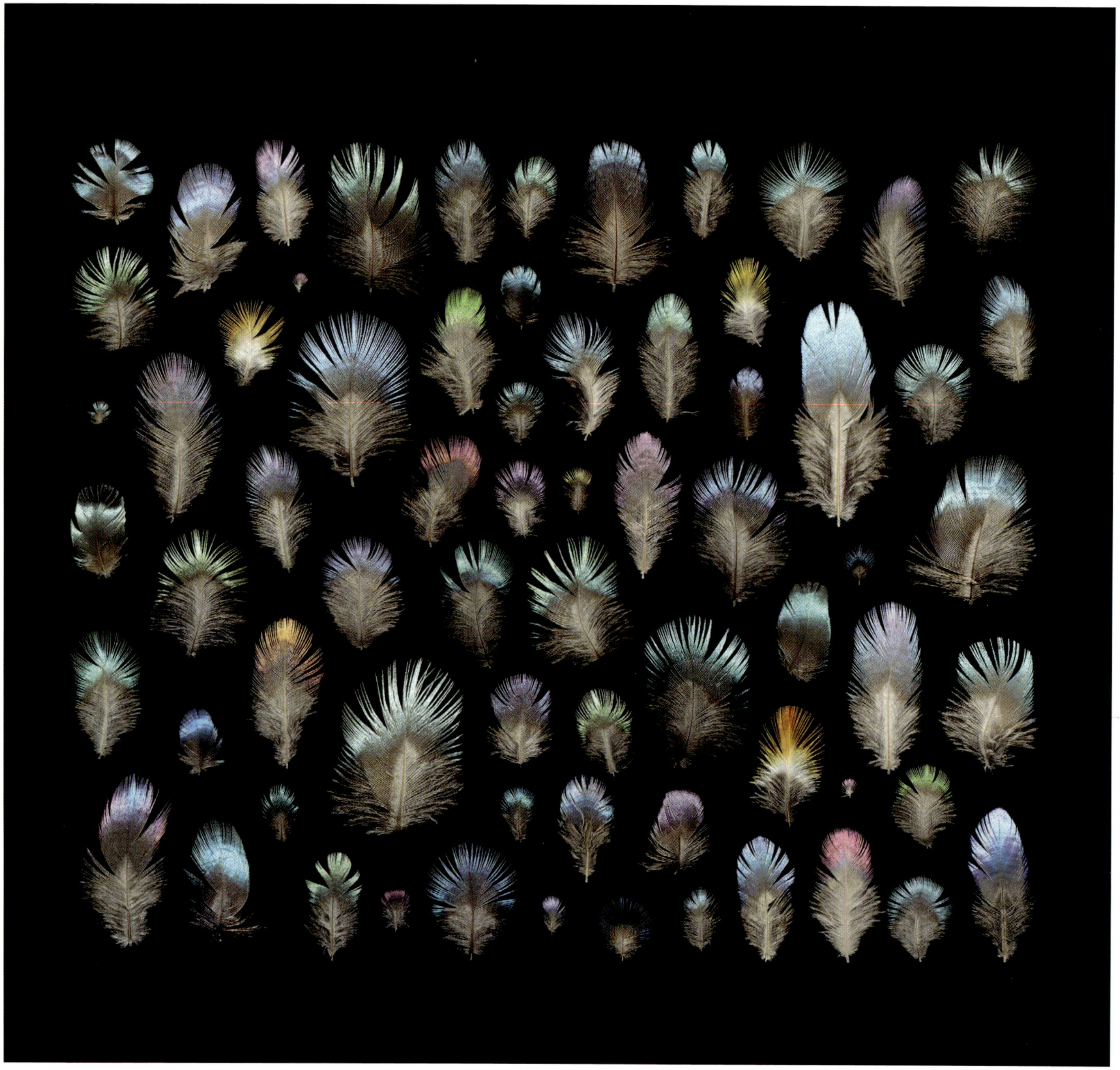

AFRICAN STARLINGS

I photograph individual feathers and combine them into single images, keeping their relative sizes. The results can be printed in large sizes and keep the details. One hundred and eight species of starlings grace the earth. Two are considered pests outside their native haunts: the European starling has been seen on antarctic islands and the common myna is expanding its range, especially in Australia.

(RIGHT)

IMPEYAN PHEASANT FEATHERS

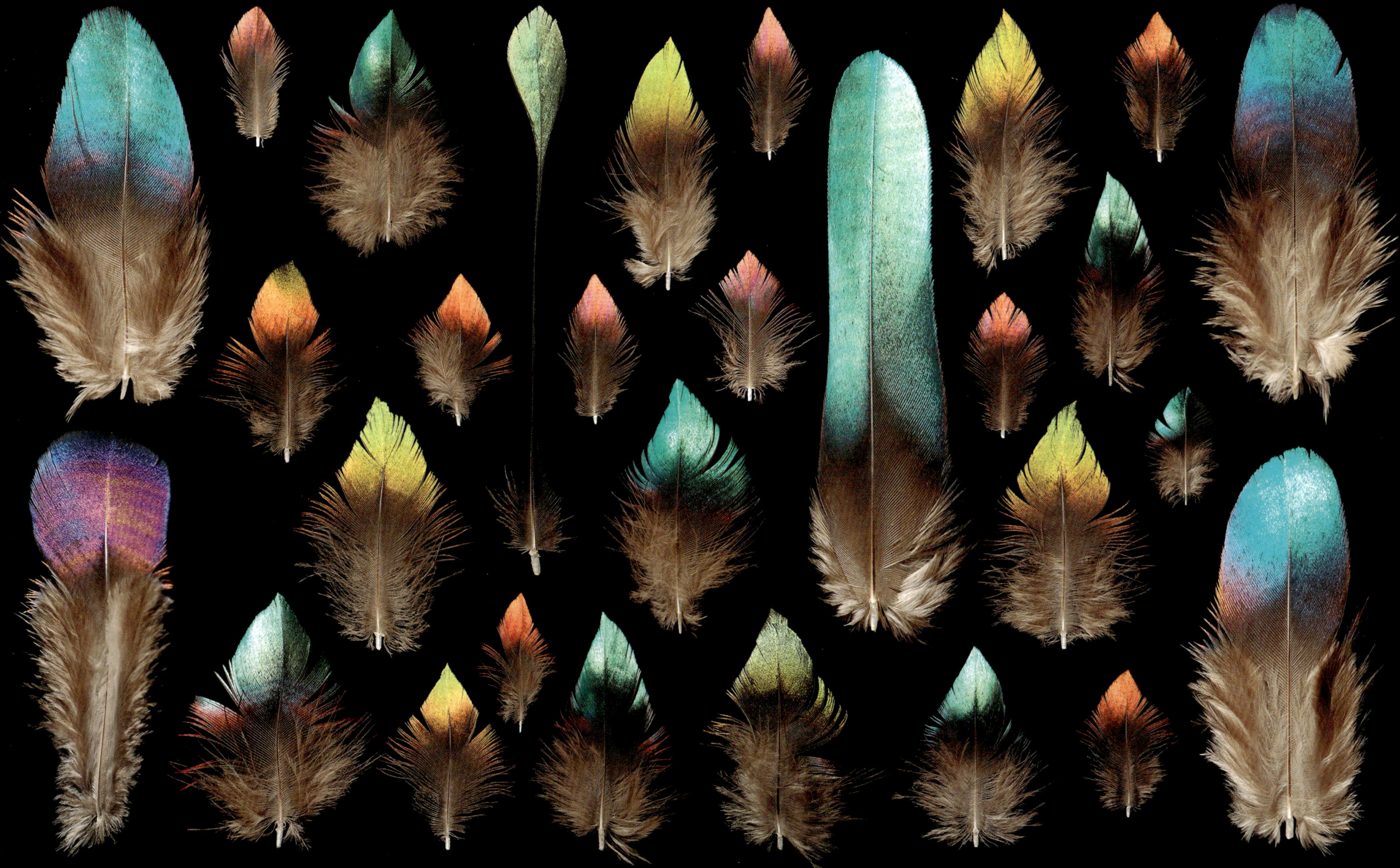

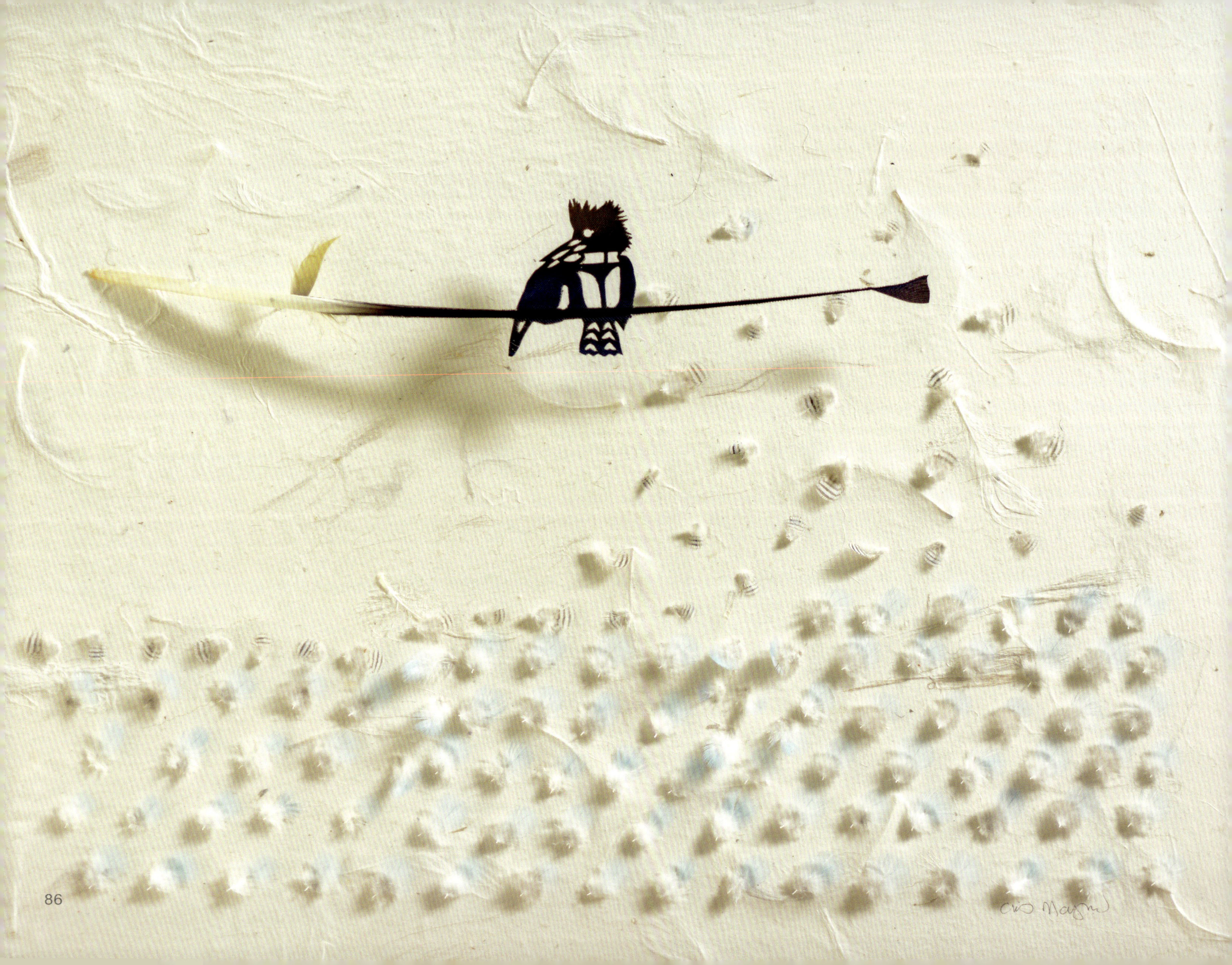

BIBLIOGRAPHY

Bashford, Dean. *The Long Feathered Fowls of Japan*. 1907. The Century Company, New York.

Brown, Roy. *Tracks and Signs of the Birds of Britain and Europe: An Identification Guide*. 1987. Helm Guides. Bloomsbury, London.

Cieslak, Marian and Boleslaw Dul. *Feathers: Identification for Bird Conservation*. 2006. Natura Publishing House, Warsaw.

Hanson, Thor. *Feathers*. 2011. Basic Books, New York.

Haupt, Lyanda. *Urban Bestiary. 2013*. Little Brown and Company, New York.

Jung, Carl Gustav. *Man and His Symbols*. 1964. Doubleday, New York.

King, Heidi. *Peruvian Featherworks: Art of the Precolumbian Era. 2013*. Metropolitan Museum of Art, New York.

Prum, R. O. 1999. *Development and Evolutionary Origin of Feathers. Journal of Experimental Zoology (Molecular and Developmental Evolution)* 285: 291-306.

Prum, R. O., and Torres, R. H. 2013. *Fourier Blues: Structural Coloration of Biological Tissues. Excursions in Harmonic Analysis*. Volume 2: 401-421 Springer, New York.

Rothenburg, David. *Survival of the Beautiful*. 2012. Bloomsbury, New York.

Schattschneider, Doris. *M.C. Escher: Visions of Symmetry*. Rev. ed. 1990. Abrams, New York.

Scott, McFarland. *Bird Feathers: A Guide to North American Species*. 2010. Stackpole Books, Mechanicsburg, Pennsylvania.

Sibley, David Allen. *Sibley's Birding Basics*. 2002. Alfred A. Knopf, New York.

Takada, M. and T. Kanochi. 2004. *The Feathers of Japanese Birds in Full Scale*. Tokyo: Bunich Sogo Shuppan, Japan.

Tennekes, Henk. 1997. *The Simple Science of Flight: From Insects to Jumbo Jets*. MIT Press, Cambridge.

Thompson, Renee. *The Plume Hunter*. 2011. Torrey House Press, Salt Lake City, Utah.

UN's Food and Agriculture Organisation (FAO) FAOSTAT 2011 world chicken production. *http://faostat.fao.org/site/573/default.aspx#ancor*

William, Chyi. *Feathers of Birds in Taiwan*. 2006. Taipei, Taiwan.

KINGFISHER PREEN
Blue & gold macaw feather and parakeet feathers
11 x 14 inches

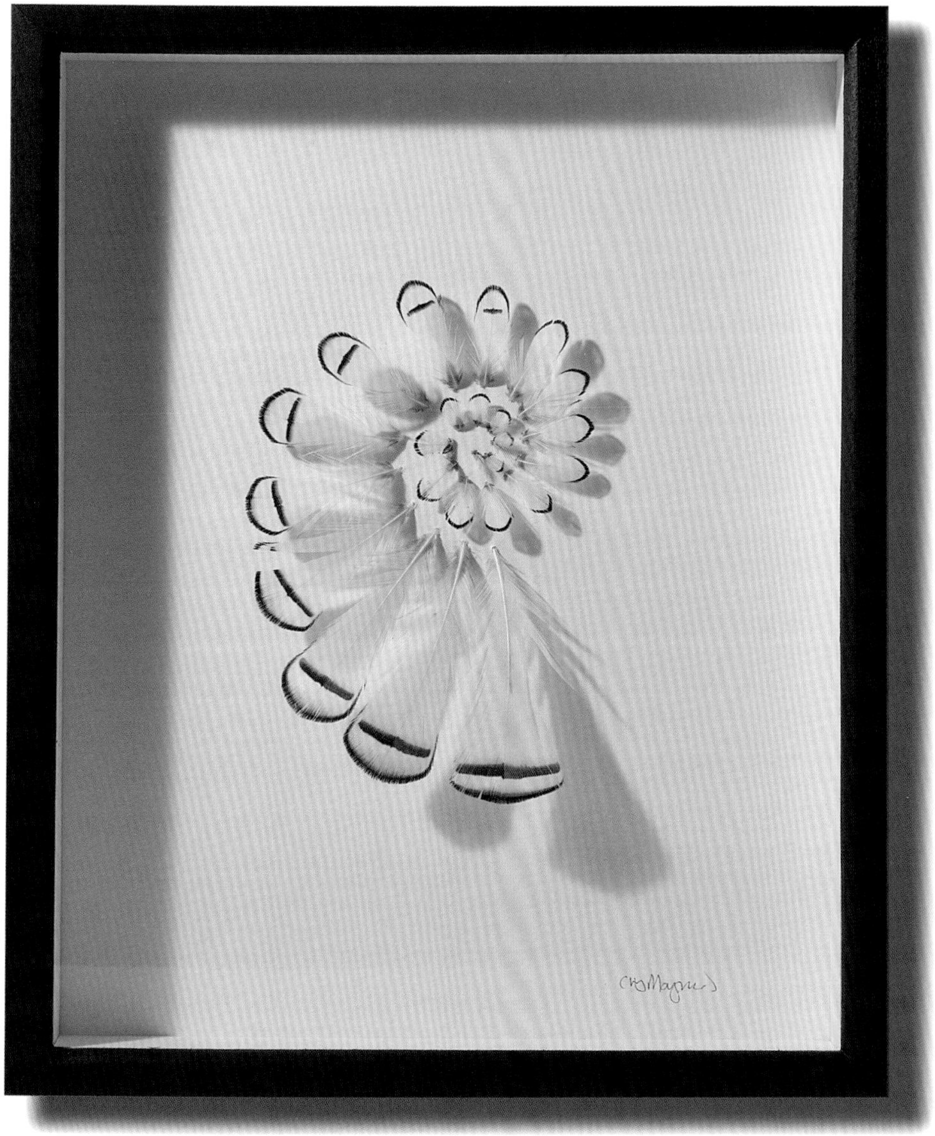

WHITE SPIRAL
Lady AMherst cape feathers
11 x 14 inches

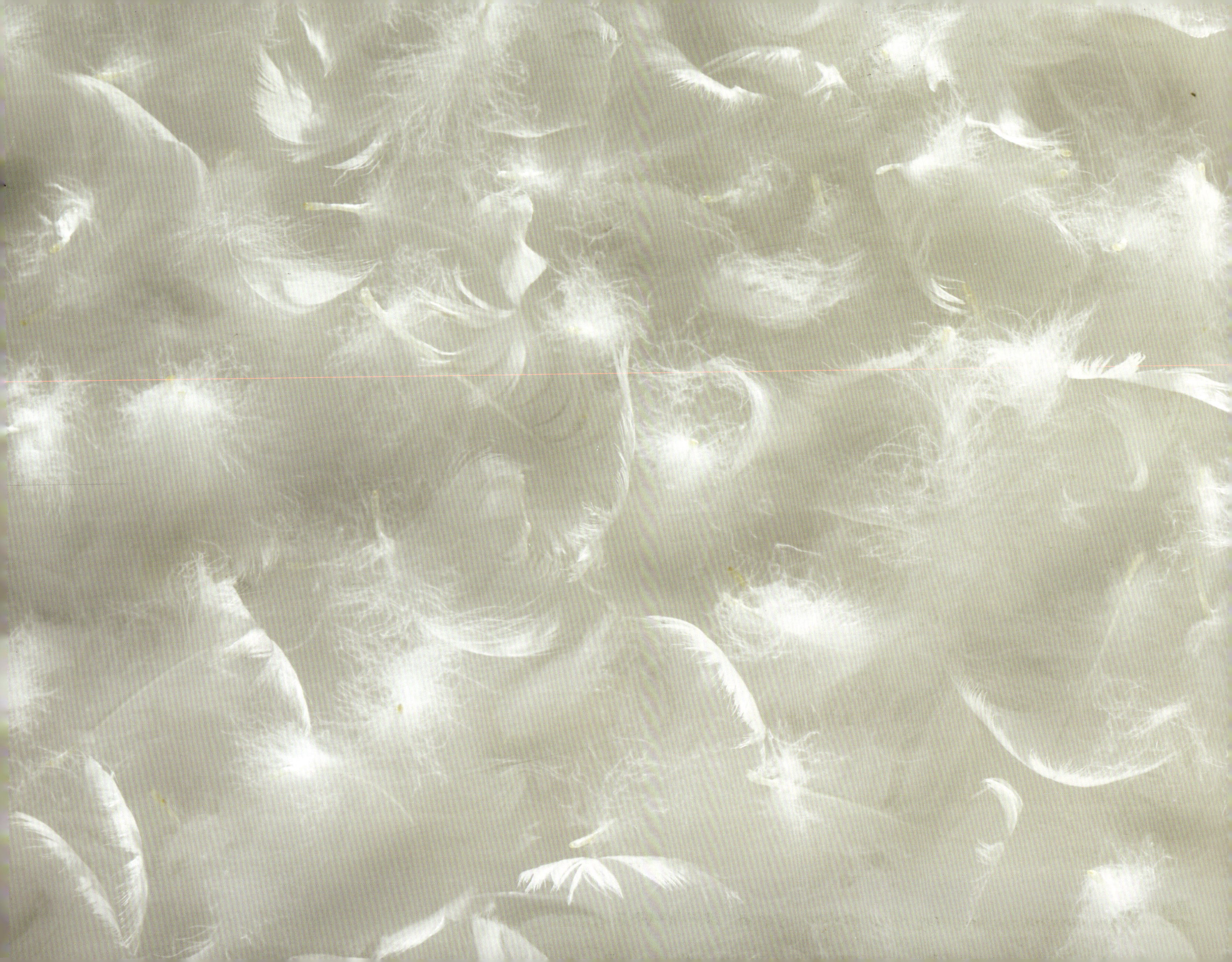